Scientific Models

Philip Gerlee · Torbjörn Lundh

Scientific Models

Red Atoms, White Lies and Black Boxes
in a Yellow Book

 Springer

Philip Gerlee
Mathematical Sciences
University of Gothenburg and Chalmers
  University of Technology
Gothenburg
Sweden

Torbjörn Lundh
Mathematical Sciences
University of Gothenburg and Chalmers
  University of Technology
Gothenburg
Sweden

ISBN 978-3-319-27079-1   ISBN 978-3-319-27081-4   (eBook)
DOI 10.1007/978-3-319-27081-4

Library of Congress Control Number: 2015957778

Printed on acid-free paper

This Springer imprint is published by SpringerNature
The registered company is Springer International Publishing AG Switzerland

Preface

As biomathematicians, we work in the borderland between different sciences. Not only between mathematics and biology, but we have also discussed scientific problems with chemists, physicists, computer scientists and medical doctors. During such discussions on scientific questions, methods and conclusions, we have on several occasions been struck by the difficulty of establishing a connection with scientists from other disciplines. An obstacle in this interdisciplinary dialogue has often been our diverging views on the concept of "scientific models". The problem was in some cases made worse by the fact that we did not recognise our differing views, and therefore did not even discuss what each participant in the project actually meant by a "model".

The purpose of this book is to avoid such confusion and to facilitate interdisciplinary communication, which these days is becoming more and more common. Our aim is not to convey and advocate a typical or consensus model within the natural sciences, but rather to show the diversity of models that exist within science. Each discipline has its own methods and tools, and since modelling (often tacitly) is central to research, it is necessary to have a comprehensive understanding of the topic if interdisciplinary work is to be successful.

Another intention with this book is to provide a basic and broad introduction to modelling and to describe how it fits into contemporary scientific practice. As such it is intended for students in all fields of natural science. We were never during our education offered this kind of comprehensive introduction to models and modelling. Instead it is something that we, like many others, have picked up in a piecemeal fashion, during courses and by reading the scientific literature. Our hope is that by offering the reader a solid introduction to the topic they will have a head start that will benefit them in the future.

Since modelling spans all areas of science it is impractical to provide an exhaustive description of the topic. Our intention is not to provide a complete philosophical analysis of the topic or to carry out an in depth historical analysis of the concept, but rather to make it accessible to researchers, students and the general public.

During the course of writing this book, we have been helped by a number of knowledgeable and generous people: Martin Nilsson-Jacobi, Helena Samuelsson, Edvin Linge, Johanna Johansson, Henrik Thorén, Bengt Hansson, and Jonatan Vasilis och Staffan Frid. Lastly, we would like to thank our editors Eva Hirpi and Olga Chiarcos, and all the scientists that we have interviewed.

Gothenburg   Philip Gerlee
Stanford   Torbjörn Lundh
14 March 2016

Contents

Prologue

On 16th September 2008, on the eve of the coming credit crisis, the Federal Reserve announced that they had created a secured credit facility of almost $85 billion in order to prevent the collapse of one of the largest insurance companies in the world, the American Insurance Group (AIG). The reason for this move, caused by an enormous deficit in liquidity, was the trade in credit default swaps (CDS) that AIG had been engaged in since the late 1990s. In the essence, CDS is an insurance which guarantees the buyer a certain compensation from the seller if a specific third party goes into bankruptcy and defaults. The initial intention of this financial instrument was as a means for companies to protect themselves against the default of corporations to which they themselves had lent money, but soon it was realised that this new financial derivative was highly lucrative and CDS were created and sold that had little to do with debt protection and more to do with financial speculation. This derivative evolved, and around 2004 AIG started selling CDS that were not designed to provide protection against a simple default, but to provide insurance against securities called collateralised debt obligations, a collection of debts such as house mortgages, car loans and credit card debts. In order to safe-guard them against the excessive risk and to price these complex derivatives, AIG used mathematical risk models to assess the probability that a payment to the buyer of the CDS had to be made. What the models did not account for, however, was that the underlying value of the collateralised debts would fall, in which case AIG was obliged to pay the buyer of the CDS additional money in the so-called collateral. This is precisely what happened in the early days of the credit crisis when the trust in collateralised debts plunged: first Goldman Sachs demanded $1.5 billion in collaterals, followed by Barclays PLC, the Royal Bank of Scotland and many others. But AIG was considered too big to bust and by early 2009 U.S. taxpayers had provided over $180 billion in government support to AIG. Using mathematical models to assess a risk was not invented by AIG; they have been in use since the 1970s. These mathematical tools are used by all major banks and investment companies, but the case of AIG shows what can happen if they do not work as expected: the losses can become enormous.

Although similar failures threaten the entire financial system, these problems are diminished when compared with what humanity faces when it comes to the climate of the planet. The causes, the magnitude and the effects of current climate change are still hot topics, and the different camps in this debate are rounding up their arguments, often, but not always, with scientific backing. How much we need to reduce our carbon dioxide emissions and how high the taxes should be on fossil fuels depend on the effects that we assume these will have on the climate. Simply put, increased emission of greenhouse gases leads to changes in the atmospheric composition that disturbs the radiation balance of the planet, which in turn gives rise to an increase in the average temperature and warmer oceans. When the temperature of the oceans increases, their ability to absorb carbon dioxide is reduced, which leads to an escalating greenhouse effect; but it also leads to increased formation of clouds, which reduces the amount of incoming radiation. The climate is affected by numerous such feedback mechanisms, the effects of which are almost impossible to comprehend. Together they create a web of interactions in which cause and effect might be difficult to distinguish. How should we scientifically approach such complex and critical questions? It is not possible to perform full-scale experiments with the atmosphere in order to study the effects of emissions. Instead an efficient approach is to study models of the climate, which can be used in order to simulate different emission scenarios and so estimate the impact on the future climate.[1] This data, together with a joint agreement on what constitutes acceptable changes in the climate, makes it possible to determine policies on emissions of greenhouse gases. This means that models of the climate to a large degree influence our political decisions, which in turn has a direct impact on our daily lives in terms of taxes on petrol, the price of electricity, etc.

In order to fully understand the ongoing political debate and current scientific inquiry, it is necessary to have a clear understanding of what models are and how they are used.

---

[1]See e.g. the collection of climate models that the UN panel on climate change, IPCC, has assembled: http://www.ipcc.ch.

Introduction

Using models as a means to investigate the world and do science became popular during the 19th century, chiefly within physics when ideas from classical mechanics were being applied to other fields within physics. The idea or concept of a model is however considerably older and derives from the Latin word *modellus*, stemming from *modulus*, a diminutive form of *modus*, meaning a small measuring device. Apart from the scientific meaning the word, "model" also has a colloquial meaning, where it refers to how something should look or how some procedure ought to be carried out. An example of this is decision or allocation models in politics. The difference between such models and scientific models is not that the latter are imprecise or subjective, but rather that they aren't simplifications of reality. In contrast to scientific models they describe an ideal state of affairs and prescribe how something should look or be carried out. The easiest way to delineate these two conceptions is to view decision models and their like as archetypes, while scientific models function as simplifications.

In the sense of a description of how something is constituted or works the word model also applies, at least to some degree, to the world views that existed before the scientific revolution of the 18th century, such as the geocentric model of the solar system. An important difference between this use and the modern concept of a model is that the scientists of antiquity and the Middle Ages didn't consider their models as simplifications or idealisation of a more complex world, but rather as direct representations of reality. In this sense the pre-modern concept of a model lies closer to models as ideals, and not models as simplifications. It is also worth mentioning that the word model has had its current meaning only since the beginning of the 20th century, and that before then it was used exclusively to denote actual physical models. What we today call scientific models have historically been denoted idealisations, abstractions or analogies. For the sake of simplicity we will resort to the modern usage even when discussing historical facts.

The two examples presented in the prologue stem from two very different parts of society, and their purposes and aims are quite different. In the case of financial derivatives, models are used in order to estimate their value in the present and a short time into the future, while models of the climate are meant to produce predictions a

© Springer International Publishing Switzerland 2016
P. Gerlee and T. Lundh, *Scientific Models*, DOI 10.1007/978-3-319-27081-4_1

hundred or so years into the future. The models also differ in their complexity and scope: financial models are meant to produce fast results (often within a second or two), while climate models describe a higher number of variables, such as temperature and humidity at a large number of locations across the planet, and therefore it can take weeks or months before an output is produced.

Despite their differences these two examples highlight the fact that scientific models are far more important than one initially might have guessed. Models are relevant not only at the frontiers of science, but are also involved in and control many of the societal functions we all take part in. Models decide to what extent, and at what interest, you are allowed to borrow money, how far you have to walk in the supermarket to buy milk, what your pension will be on the day you stop working, and for how long you have to wait at the traffic lights when driving your car. In short, we are surrounded by models.

Apart from the fact that models are applied in such varying circumstances, they can also be of many different kinds: mathematical models that help us predict the weather, the climate or pollution from industry; scale models that make it possible to estimate the lifting force of an aeroplane wing or illustrate how the atoms are placed in a molecule; a diagram that shows how decisions are taken within a multi-national corporation or how different species interact within an ecosystem; or animal models that can help us understand the mode of action for a new drug. The list can be made even longer, and every scientific discipline, be it meteorology, economics or ecology, has its own way of developing, using and relating to models in its quest for knowledge. The common denominator for this wide spectrum of models is that they represent simplifications of reality that offer access to and an understanding of distinct aspects of the world.

This brief description might trivialise the concept, but we hope to show in this book how multi-faceted and interesting the concept of a model really is. To achieve this we will give an overview of the history of models and modelling in different disciplines. The book also contains a discussion on philosophical aspect of modelling, such as the relation between model, theory and experiment, and also a section on how the accuracy of a model should be viewed and assessed. In order to connect with the actual use of models in modern science, we will present the results of a qualitative interview study that investigates the views on models and modelling among a number of scientists. Apart from giving a basic and inter-disciplinary account of models and their use in science, the book also aims to bridge the gap between different disciplines and foster an understanding of different approaches to modelling, and tries to do this by clarifying differences and similarities in the use of models. Since the concept of a model permeates almost all scientific work our account will be far from complete. This book is not a complete guide to scientific modelling, but should rather be viewed as an initial attempt at conveying a coherent understanding of the topic.

Models as Illustrations and Illustrations as Models

Before we begin a thorough analysis of scientific models and their function in science, we will try to convey a feeling for what models are and what they can achieve. We would like to show how models can be used as illustrations of a phenomenon, or rather a subset of a phenomenon, but also that the reverse is true, that illustrations can function as models.

Assume that we are studying a colony of *E. coli* bacteria that grows on an agar plate, a flat plastic dish filled with a gelatinous solution of the sugar agar. If we further assume that the population of bacteria is small at the start of our experiment and that the supply of nutrients in the dish is sufficient, then we can safely assume that each bacteria divides as often as possible, which results in a doubling of each bacteria on a regular basis. The growth rate, i.e. the number of new bacteria per time unit, therefore becomes proportional to the current number of bacteria in the population.

How do we best illustrate these assumptions and the dynamics that follow? Let us consider a number of variants:

The number of bacteria double each generation. This is just shorthand for the above assumptions.

As an illustration see Fig. 1, where time passes to the right and the bacteria and their offspring are represented as circles.

The number of bacteria grows exponentially. This apparently simple and colloquial description is actually based on the solution of a differential equation.[1]

$n_{i+1} = n_i + n_i$, **where n_i is the number of bacteria after i generations**. This description is merely a translation of the first one into a mathematical/algorithmic language.

All the above descriptions convey the same message, some more clearly than others, but in essence they are the same. Which one would you prefer? Which one is the most illustrative? That of course depends on your preferences, what the purpose of the illustration is, the background of our audience etc. For example in a popular science context it would be preferable to chose the phrase "The number of bacteria double each generation" or perhaps the figure.

Now returning to what we initially discussed, the growth of the bacterial colony on the agar plate, we realise from the above illustrations that the situation is untenable in the long run. The population cannot continue to grow exponentially indefinitely,

---

[1] This can be seen if we start with the first description and translate it into discrete mathematical notation as in the last case. We then turn this discrete description into a continuous one, by introducing a continuous time variable t and a growth rate β. We can now write $n(t + \Delta t) = n(t) + \beta \Delta t n(t)$, where Δt is some small time interval. The next step is a transition to a continuous equation by letting Δt tend to zero, which in the limit yields the differential equation $n'(t) = \beta n(t)$. This equation has the exponential equation $n(t) = n(0)e^{\beta t}$ as a solution, where $n(0)$ is the number of bacteria at the start of the experiment.

Fig. 1 An illustration or model of bacterial growth at optimal conditions. Since all bacteria in the population divide at regular intervals (a couple of hours) the total number of bacteria doubles at each generation. This gives rise to exponential growth that can also be described mathematically with the expression $n(t) \sim e^{\beta t}$, where β is the growth rate that depends on the generation time

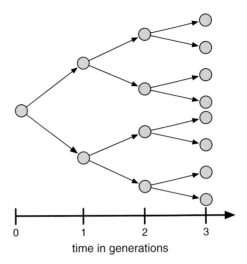

time in generations

since no matter how large the plate is, space and nutrients are limited.[2] The model thus has an obvious limitation, it only applies to the early stages of colony growth. If we want to be able to describe the long-term dynamics we need to change our assumptions and include some sort of competition for space and nutrients.[3]

What the above examples show is that simple models can be viewed as illustrations of a system or process, and that there are a number of ways to illustrate a particular system. Which one is most suitable depends on the context and audience.

We now turn the tables and ask ourselves if an illustration, as in a drawing or picture, can function as a model. In fact it can. A good illustration fulfils many of the criteria normally put on models, in particular when it comes to conceptual models that often serve as way of arranging our thoughts around a certain system, process or topic.

An example of this are the illustrations used in books of flora. These figures are not photographs of real plants, but hand-drawn pictures of typical specimens. Why is this? One might expect that an exact representation of real plants would be superior when it comes to matching what is shown in the book to what we see in nature. For example, wouldn't it be easier to tell the difference between an edible and a poisonous mushroom from a photograph of the two? In fact the answer to this question is no, and the reason is that an artist's rendition, similarly to a modeller, has the advantage

---

[2]This is of course also true for us humans on our finite planet. Thomas Malthus raised this concern in 1798, and claimed that since human populations grow exponentially while food production only grow linearly we are heading for a disaster. This phenomena has been termed the Malthusian catastrophe.

[3]Such a model can be viewed as an extension of simple exponential growth, and can be described in terms of a differential equation $n'(t) = \beta n(t)(1 - n(t)/L)$, which was first written down by the Belgian mathematician Verhulst in 1838. It is known as the logistic equation, and contains the additional parameter L, the carrying capacity, which represents the maximal number of bacteria that the agar plate can sustain.

Fig. 2 Two pictures of chanterelles. The *top* picture shows a realistic photograph, while the *lower* one is painted with watercolours. Which one is better? The answer naturally depends on context and purpose, just as in the case of bacterial growth that was discussed on p. 3. *Upper panel*: Published with kind permission of © Michael Krikorev. *Lower panel*: Eugen Gramberg, *Pilze unserer Heimat* (1913)

of exaggerating characteristic properties of a species while suppressing those that are of minor importance (see Fig. 2). In this sense there are many similarities between an artist depicting flora and someone devising scientific models. Both of them start by observing many different specimens of the species (instances of a system) and summarise the findings in a picture (a model) that is meant to capture the essence of the species (or the system) without becoming simply an average of all specimens. This process requires many sketches and drafts, and also communication with scientists who are familiar with the species (system). Creating a model can therefore, from this perspective, be viewed as a craft or even a fine art.

Three Routes to Modelling

After this initial, and hopefully inspiring, take on models we will now try to close in on the topic of modelling from a intuitive point of view. We will bring this about by discussing models in three separate contexts: mental/cognitive, artistic and scientific. Our hope is that this will bring the reader closer to the subject and provide a deeper understanding, which in the end hopefully will make it easier to appreciate

the similarities between modelling in different scientific disciplines. The presentation
can also be viewed as a progression from informal mental models, which guide our
thoughts and never leave our mental life, to artistic models, where conventions and
rules control the ways models are formed and applied, and finally scientific models
that are subject to stricter demands, such as rationality, empiricism and objectivity;
principles that apply to all scientific work.

Models in Our Minds

In the introduction to this chapter we mentioned that a common feature of all models
is that they are simplifications or reductions of more complex real-world phenomena.
Certain aspects are stripped away in favour of those that are viewed as interesting and
more general (see Fig. 3). This is a process of generalisation, from the concrete to the
abstract, but sometimes also the reverse. A mathematical model is more abstract than
the reality it tries to describe, while plastic models of atoms and molecules are more
tangible and concrete than the microscopic reality they are depicting, which for our
senses is imperceptible. However, in both cases it is a matter of simplification. The
model only retains a minority of the defining features of the real object. By doing this
the scientist can safely state that the model is describing a system of a certain kind or
class, not a specific instance of the system. For example, the scientist creates a model
for the progression of a certain disease, not the disease progression in a specific
patient.[4] In much the same way the engineer uses a mathematical model of fluid flow
in order to describe the drag of the hull of a ship without taking into consideration
the colour of the hull or the salinity of the water. What is being ignored in each
case is highly dependent on the system under consideration and also depends on the
type of questions that are being asked, and in addition is affected by the subjective
preferences of the modeller.

This process of simplifying the object of study might seem trivial, but does so
simply because it is a fundamentally human activity. We are constantly creating
generalised conceptions about the outside world in order to organise the constant
flow of sensory perception we are exposed to. Let us illustrate this with an example.
If you happen to see an object consisting of a wide board that balances on four sticks,
then you'd most likely instantly identify it as a table. Naturally the dimensions of the
"table" will influence your classification, as well as the context you see it in. Assuming
that you don't have any particular interest in that specific table, a reasonable level of
description would be simply "a table", perhaps with a few additional adjectives such
as "small" or "black". A more thorough description of how the legs are assembled
or the respective angles would probably not interest you, at least not initially. The
concrete manifestation of the table, in terms of visual input, has via mental processing
been converted into an abstract symbol that represents the real object. Along with

---

[4]Although it is believed that treatment might be improved if given on a personal basis, and for this
we need "personalised" models.

Fig. 3 The painting *La trahison des images* (1929) by the artist René Magritte, © René Magritte/Bildupphovsrätt 2016. With this painting the artist examines the relation between reality and image. The viewer is well aware that it isn't a real pipe that is shown; despite this the subtitle, which in English reads *this is not a pipe*, is surprising and somewhat comical. The painting highlights an important property of scientific models; they are never exact replicas of the system we are hoping to understand. Instead they are simplified with a specific purpose in mind. The pipe in the painting is depicted in order to *look* like a real pipe, but it differs in most of the other qualities a real pipe harbours such as mass, smell and texture

this representation comes a number of tacit assumptions about the object: in the case of a table we probably assume that it can hold a certain weight, and that the legs of the table have the same length. This symbol is then utilised in order to reason about the object and its relation to other objects. What happens if I place this book on the table? Will the table collapse?

Such a process is however only possible if the person in question possesses a mental archetype of an ideal table, from which the symbol of the current table is generated. In this mental process the specific table is represented with the help of a general notion of a table, and it is these general concepts we refer to as *mental models*. They are the simplified mental conceptions that we use in order to identify objects that in reality are very complicated and hold a lot of details. Or to put it another way, we apply mental models to real objects and in this process, in an effort to make the world understandable, simplify and discard details. These models are static representations of real objects, that closely resemble corresponding models in science, such as material and iconic models, but do we also have mental counterparts of the large class of dynamic and time-dependent models used in science?

Imagine a situation in which you need to run a couple of errands in town. You look through the window and notice a slight drizzle. Is the rain heavy enough to warrant an umbrella? By creating a visual image of how wet your coat and shoes will become you decide that taking an umbrella is probably a good idea. But just as you're about to leave you take one last look out of the window and notice that it's getting brighter. Maybe the rain will stop soon? You reassess the situation, taking the new facts into consideration. In doing so you have tried to make a prediction about the future with the help of existing knowledge. However, you only made use of a subset of all available facts; e.g. you refrained from checking whether the air pressure was rising or falling, or what the forecast said about the weather that afternoon. In conclusion you have, during a brief moment, created an extremely simplified and limited model in order to make an everyday decision. This is just a single example, and it's easy to see that we continuously make use of both static and dynamic models in order to navigate in our surroundings. They present an indispensable tool for representing and structuring our sensory perceptions, and based on those we make decisions about how to react.

In the above example, which is more complicated than the static model discussed initially, several symbols that represent parts of reality interact with one another and produce an outcome. For example one needs to have a notion of how the rain "interacts" with the coat, i.e. how fast will it get wet under varying conditions such as the wind speed and the intensity of the rain? In this type of reasoning we can anticipate a connection to scientific models, in which typically several objects or quantities interact with each other through a set of mechanisms that are controlled by the parameters of the model.

But what is the difference between mental and scientific models? We can start by noticing that humans rarely are rational in the assessment of their surroundings. If I went out when the rain was pouring down yesterday, then the likelihood of taking an umbrella is probably higher than if we stayed in, although we don't have any rational reason to do so. Scientific models are on the other hand subject to the rules of logic and rational deduction. Mental models are also in a trivial way subjective, since they always involve a subject, namely the person forming the mental model and who reasons with it. In this sense they don't exist independently of someone's mental life. Science on the other hand always strives for objectivity, or at least inter-subjectivity, i.e. knowledge that is valid for everyone. Despite these differences we would still like to argue that the similarities are many, and in one way this is natural, and even trivial. Scientific models are the creation of man, and to use mental models as the basis (together with empiricism, rationality and objectivity, the standard requirements of modern science) can in fact be said to be the only option.

Models as Caricatures

We can also approach models from a slightly less abstract angle by viewing them from an artistic perspective. What we are referring to are not concrete models made of stone or bronze, nor models as in ideals (as in the phrase "to model for something"), but rather the process or technique with which the artist represents a real object in a composition. A striking example of this is the way a caricaturist depicts people in such a way that certain features and characteristics are exaggerated while others are neglected. This was alluded to above in the discussion of the chantarelle drawing, but in the case of caricatures it becomes even more pronounced. The difference between an "exact" photographic representation and a caricature is shown in Fig. 4, which

Fig. 4 Photograph and
caricature of August
Strindberg. A good
caricature must, just like a
good model, contain the
essential features, but not
excessive detail. The
caricature was drawn by Carl
Larsson (1884)

depicts the Swedish writer August Strindberg in two completely different ways. The artist has focused on a small number of characteristic features, such as Strindberg's hair, and has been very economical when it comes to the general level of detail in the picture. Despite this one can easily recognise the person in the drawing, which reveals the skill of the artist. The ability to achieve this high level of similarity can be said to be two-fold: firstly, one has to be able to identify the characteristic features of a person, and secondly one has to be able to realise them in a convincing manner. If we make a parallel to scientific models this two-fold knowledge reflects the ability to identify the processes and mechanisms that are of importance in a system, and secondly to represent them in a correct way within the model.

One could argue that a photographic rendition of an object or person is the most "accurate". Such an argument relies partly on the fact that our visual sensation is the dominant mode of perception, and this is difficult to argue against, but this view lacks an appreciation of the fact that it is not the uninterpreted "photographic" impression that we perceive and remember, but an interpreted and processed one, coloured

Fig. 5 Magritte's painting
La Condition Humaine
(1933), © René
Magritte/Bildupphovsrätt
2016. This piece can be
viewed as a reflection on the
arts and the reality it is trying
to depict. Is an exact copy
the most desirable method,
or are there better ways of
representing a scene?

by feelings and associations to the objects and features the photographic image contains. If an artist experiences a certain feeling or atmosphere when confronted with a scene, then that feeling is not necessarily best conveyed by a photo, but rather by a representation in which feeling can be expressed and details highlighted. It is worth noting that even photographers make use of various tricks and technical devices, such as lighting, depth of field and different exposure times, in order to accentuate features and create certain moods. The most realistic depiction is rarely, or possibly never, the best way of representing a person, place or mood (see Fig. 5). Or in the words of scientist, the most realistic model is rarely, or possibly never, the best way to describe a given phenomenon. On the contrary it is models that neglect certain aspects of reality and focus on other aspects that give the most insight into the phenomenon.

Models in Science

Scientific models can be said to resemble both mental and artistic models by being both abstract and concrete. Mathematical and theoretical models can be likened to mental constructions, but in contrast to their mental counterparts they are subject to strict and formal rules when it comes to manipulation and interpretation. Scale models and schematic models, on the other hand, have, in being material, more in common with artistic models. But if scientific models cannot be isolated on the spectrum that ranges from concrete to abstract, what is it then that sets them apart from other types of simplifications, generalisations and abstractions?

A possible answer to this question is that they satisfy the requirements of rationality, empiricism, and objectivity that other scientific methods and tools are usually subject to. Models should not build on unfounded assumptions (they should obey rationality), they should in one way or another be in agreement with reality (be empiric), and finally the results and outcomes of the model should not depend on who uses or evaluates it (be objective).[5]

This last requirement is also connected to an important property of models that might seem incongruent with our common view of science as a means of establishing true facts about the world, but that makes perfect sense when we view scientific models together with mental models and artistic representation. Namely the fact that scientific models are always wrong. This is trivially true, since models per definition are simplifications of reality, but from another point of view it is interesting since scientific models have turned out to be useful tools in a wide range of scientific disciplines. How can faulty models provide us with knowledge about the world? Our mental models are obviously indispensable for us when it comes to interpreting and navigating reality, but from this observation it doesn't follow that a similar strategy would be successful in our scientific endeavours. We don't aim for a complete answer to this difficult question, and instead hope to illuminate it from different perspectives by discussing the historical roots of scientific models, their structure and how they are being used in different scientific disciplines.

In fact scientific models play a central role in all branches of science.[6] Excluding models would leave little or no science behind. This might not be obvious at first, and depends to a large extent on the fact that models and modelling appear in so many different guises. A physicist would probably see few similarities between her mathematical models, which describe reality using abstract mathematics, and the animal models of a life scientist, in which human diseases can be induced and studied (in e.g. mice that are far easier to study, particularly from an ethical perspective). Of course these models are quite different, but on closer inspection one realises that the points of contact are numerous (which often can be hard to realise for the respective scientists). In both disciplines models are used in order to test hypotheses and interrogate possible mechanisms, and also form a central part of the scientific work by providing a tool for scientists to reason about phenomena. One could even say that almost all scientific work amounts to constructing, verifying and reasoning with the aid of scientific models.

To get a proper view of the scientific work carried out in other disciplines it is instrumental to know what kind of models exist, how they are used, and also their limitations. This is particularly relevant in the current scientific community where

---

[5]This requirement might be a bit too optimistic, since often the scientific debate concerns different opinions and interpretations of a single model.

[6]Except paradoxically one branch, see p. 57.

interdisciplinarity is becoming more and more common. In order for scientists to be able to reach across disciplinary boundaries it is imperative to avoid the misunderstanding that talk of models and modelling often leads to, and this is easily achieved by having a better understanding of models from a general perspective. One way to reach that goal is by acquiring knowledge about the origin of models, the wide spectrum of existing models, and how they are applied in different disciplines. These are important matters, but before we address them we will deal with an even more fundamental question: Why do we use models in the first place?

Why Do We Use Models?

The straightforward, but not particularly illuminating answer to this question is: because they work. Models help us gain knowledge about a reality which often is complicated and difficult to grasp. But this short answer is far from satisfactory, and we will in this section try to give a more variegated answer to the above question. Hopefully the message will become even clearer further on in the book, by which time the reader will have gained a deeper understanding of how models are applied and used in different disciplines.

The fact that models are useful can hardly be disputed. Despite the fact that they often consist of a highly simplified and sometimes distorted version of reality, they somehow convey knowledge about how the world is constituted. Exactly how this process of gaining knowledge about the world occurs through the help of inaccurate models is far from understood.[7] An indication can, however, be found in the fact that models often have overlapping functions: there are often many models that describe the same phenomenon, or part of a phenomenon, and by giving partial explanations (that taken separately are insufficient) they provide us with an aggregate understanding, which otherwise could not be reached.

The problem of understanding the world around us arises partly from the fact that it is composed of a complex muddle of processes on different temporal and spatial scales. This makes it hard for the unaided human intellect to grasp. For example weather systems move across hundreds of kilometres, while the constituent clouds only measure a couple of kilometres across, and the droplets forming the clouds are of millimetre size. Mountain ranges are worn down during millions of years, while new islands can be formed during a volcanic eruption that only lasts a couple of hours. Despite these vast differences in spatial and temporal scales the fundamental laws that dictate the dynamics of the different processes are identical.

In order to understand different aspects of a complex phenomenon such as weather, it's a reasonable strategy to observe and study different parts of the system as separate

---

[7]One could actually turn the argument around and claim that models, by being gross simplifications, provide us with a view of reality which is much clearer and devoid of the distracting details that we normally perceive. However, this still doesn't explain why models are useful, but hints at the fact that regularity and generality are salient features of the world.

entities, i.e. divide the phenomenon into parts of appropriate "size", and then study each part in isolation.[8] For example it might be a good idea to first understand the dynamics of a single drop of water before one tries to construct models of how droplets interact and finally attempt to explain how clouds, which contain billions of water drops, behave. By merging our understanding of the different subsystems we can thus acquire a more fundamental understanding of the entire phenomenon. This division into less complicated processes and building blocks can be viewed as the first step in constructing a model and often corresponds to an intuitive subdivision we carry out as humans when perceiving a complicated phenomenon. In the process of isolating subsystems we create distinct boundaries that provide a basis for further study of the system. This approach, commonly termed bottom-up modelling, is often successful, but in many cases it is impossible to link all the involved scales and sub-phenomena in a quantitative fashion. For example we are still lacking a fundamental understanding of how clouds are formed and how they impact the climate. In such cases one needs to approach the phenomenon top-down, where fundamental levels of the system are ignored.[9] Knowledge of the basic building blocks and laws of nature that govern the phenomenon is, however, no guarantee for understanding its dynamics and properties.[10] Interactions and relationships are often highly complex, and if the dynamics can be formulated in terms of mathematical equations, then these are in most cases impossible to solve analytically. In order to make progress we need to simplify the phenomenon. For example, in order to understand the properties of a large molecule, such as a protein, it is intractable to write down the quantum mechanical equations and solve them exactly. Instead we make assumptions and simplifications concerning the interactions of the atoms that constitute the molecule.

The simplifications that are reasonable and useful are however not easy to arrive at, neither is knowledge about the parts of a phenomenon that safely can be ignored. This requires insight into the structure of the system and also previous experience of similar systems. In addition it depends on the objectives of the study, and on what aspects of the system are considered relevant. It might also be the case that exact knowledge about mechanisms and processes acting within the system is lacking. In cell biology new mechanisms by which human cells regulate their behaviour are continuously being discovered, but this does not stop scientists from building imperfect models. Independent of our current knowledge of a system, in order to approach the problem at all, one needs to reduce the complexity.

We can from this short discussion identify two distinct steps in the procedure: isolation and simplification. First the phenomenon is restricted to something that is graspable, and then it is simplified into a manageable concept. One could say that a so called conceptual model of the phenomenon has been created. A conception of the extent of the phenomenon, and the mechanisms and processes acting within

---

[8]This is related to the idea that science should "carve nature at its joints", a thought that was brought forward for the first time by Plato in the dialogue Pheadrus (265d–266a), written circa 370 BC.

[9]For a discussion about these two approaches to modelling see the interviews (p. 56).

[10]If this were the case then all science would be deducible from quantum mechanics, that describes the physics of fundamental particles, which is clearly not the case.

it has been formed. This basic construction can then be used as a blueprint for further model building, something we will look closer at later in the book. But even before a concrete or mathematical model has been formulated, this conceptual model can provide insight and understanding of the phenomenon. It provides us with a possibility to start to reason and speculate about observations that have been made, and possible future experiments. A platform for twisting, turning and speculating about the phenomenon has been constructed. This preliminary stage that precedes a more sophisticated model can therefore be viewed as an indispensable part of the scientific method and a means of acquiring knowledge about the world.

Another reason for using models is of a more practical nature, and related to the cost of and the possibilities of carrying out experiments on the system under study. This is particularly important in engineering where construction of e.g. roads, bridges and vehicles is a common activity. In these cases, due to economic limitations, it is practically impossible to test every possible construction in full scale, and instead computational and scale models play an important role. Here, a model functions less as a tool for understanding, and more as an efficient tool in the process of constructing a bridge or a ship. In some cases when experiments are not only limited by economic constraints but also by ethical considerations, modelling plays an even more crucial role, for example when one needs to estimate the environmental impact of a nuclear meltdown, or the effect of an asteroid impact on Earth. Similar considerations have prompted the use of animal models in the medical sciences, since the experiments necessary to test hypotheses and possible drugs are considered unethical if carried out on human subjects. In these cases modelling is the only viable path.

Yet another reason for the success of models is their ability to serve both as vessels of knowledge and as tools for acquiring new knowledge about the world. What we mean by this is that the behaviour or properties of a given system can be represented in a simple way by using a model. It provides a compact description that has the ability to explain the structure of the phenomenon in a pedagogical way. From this perspective one might view a model as a recipe for generating the dynamics of a system, which is more useful than a long list simply enumerating all the possible behaviours of the system. On the other hand models also serve as a means of testing hypothetical mechanisms and processes within systems that have not been fully explored. By building a variety of different models of the same phenomenon it might be possible to decide which interactions and processes are acting within the system, and this might lead to the formulation of a novel theory of the phenomenon. In this instance models have an exploratory function, acting as tools with which we explore the world.

We conclude this section by recapitulating that scientific models are the basic tools used for isolating, simplifying and reasoning about natural phenomena. They also facilitate the practical management of certain questions, and they function as simple representations of acquired knowledge. Taken together this implies that scientific models are the centrepieces of the scientific method, with which we gather knowledge about the world, and they are also indispensable tools in our endeavour for novel technical solutions and constructions. That is why we use them.

History

In the previous chapter we gave an introductory explanation as to why models play such a central role in contemporary science, and also alluded to the fact that without models there wouldn't be much science left to consider. In order to fully appreciate this deep connection between science and the practice of modelling it will be useful to view models from a historical perspective. This chapter will give an account of the historical development of the concept which stretches from the scientific revolution of the 17th century to the scientific inquiry of today. In order to fully appreciate this account we will also investigate the concept of a "mechanism", which since its inception into the natural philosophy of the 17th century has become a central term in modern science. Finally, we will look closer at an example from the history of Swedish science, namely the pioneering work of the engineer and scientist Christopher Polhem, and in particular his work on scale models.

It Started as an Analogy

The first steps towards the practice of modelling, as we know it today, were taken during the scientific revolution of the 17th and 18th centuries, a time when religious world-views based on Christian belief, and ideas from antiquity, were displaced by novel ideas based on observation and critical thinking. During this period of time the empirical scientific method, based on experience, was adopted by many scientists and natural philosophers, and paved the way for many scientific achievements such as the heliocentric worldview put forward by Copernicus, the laws of planetary motion first described by Kepler, and perhaps the most important, Newtonian mechanics. These advances were strongly coupled to a mathematical approach to mechanical phenomena, and even required the invention of new kinds of mathematics, such as the differential calculus. However, describing Nature in these novel mathematical terms required both abstraction and simplification. When Galileo, and later Newton, derived

© Springer International Publishing Switzerland 2016
P. Gerlee and T. Lundh, *Scientific Models*, DOI 10.1007/978-3-319-27081-4_2

their laws of motion they made use of numerous simplifying assumptions, such as point masses and frictionless planes, which allowed for a mathematical treatment, but at the same time separated the analysed phenomenon from its real counterpart. They were both however well aware of the importance and ramifications of such simplifications, a fact which becomes clear in the following passage by Galileo which discusses the motion of falling bodies:

> As to the perturbation arising from the resistance of the medium this is more considerable and does not, on account of its manifold forms, submit to fixed laws and exact description.[1]

In the previous chapter we argued that mental models are the basis on top of which scientific ones are built, and even if this is the case, the abstraction enforced by mathematical tools and reasoning can be viewed as a first step towards the modern conception of a model. In order to describe mechanical phenomena in mathematical terms they had to be simplified to an appropriate level of description, at which stage the mathematical tools could be applied. This allowed for an accurate description of many phenomena, such as planetary motion, but the process of simplification itself also turned out to be a powerful (and general) tool that would help scientists to reach a deeper understanding of the natural world.

The school of thought which paved the way for many of the scientific achievements during this period is known as the mechanistic (or corpuscular) philosophy and among its proponents one could find philosophers such as René Descartes, Francis Bacon and Pierre Gassendi. The basic tenet of this philosophy was to view the entire universe as a machine, often in analogy with a mechanical clock.[2] Animals and plants were considered as mere machines made of flesh, and all natural phenomena were considered as explainable in terms of "micro-mechanical" actions. This was in stark contrast with the, at that time, dominating Aristotelian worldview where intention and purpose were the means by which Nature was to be explained. For example the force of gravity was, in the Aristotelian tradition, thought to emerge from material objects striving to reach their natural place in the universe. Man and in particular his soul was however excluded from the new mechanistic explanations, simply by placing them outside of the material world.[3]

The success that was achieved when mathematical and in particular mechanistic thinking was applied to classical mechanics, led to a rapid spread of this new way of investigating natural phenomena to other branches of physics such as optics and fluid dynamics,[4] that previously had been dominated by purely experimental work.

---

[1] Galileo Galilei, *Discorsi e Dimostrazioni Matematiche, intorno a due nuove scienze* (1638).

[2] An example of this analogy is provided in the following quote by Kepler: "I am occupied with investigating the physical causes. The aim of this work is to show that the machine called the Universe does not resemble a deity but a clock".

[3] This famous move was carried out by René Descartes who in his book *Les passions de l'âme* (1649) claimed that man was made out of two distinct substances *res cogitans* (mental substance) and *res extensa* (corporeal or extended substance) and that the communication between the two occurs in the pineal gland of the brain.

[4] Important works of this time are Newton's *Opticks* (1704) and Daniel Bernoulli's *Hydrodynamica* (1738).

The development of sophisticated mathematics in conjunction with the mechanical philosophy lead to a formalisation of physics as a science. This change later became known as the "mathematisation" of physics and transformed a discipline dominated by experimental work into one where mathematics and formalism played an integral part, a trend which has persisted until today. An interesting consequence of this, which didn't become obvious until the beginning of the 20th century when quantum mechanics was formulated, was that certain branches of physics had become abstract and mathematical to such a degree that the topic of inquiry was incomprehensible without mathematics, and hence the differences between a mathematical model, pure mathematics and physics had become indistinguishable. An example of this is string theory, which describes the smallest constituents of matter, not as point-like particles, but instead as vibrating one-dimensional strings. This assumption is not based on any observational data, and the theory has not been able to provide predictions that are experimentally testable. One could therefore say that string theory exists somewhere in the twilight between mathematical fiction and physical reality.

When the use of mathematics during the 18th century spread from classical mechanics into other disciplines, one of the first disciplines to adopt this new way of thinking was electrical theory. This resulted among other things in the formulation of Coulomb's Law, which describes the forces between two electrically charged particles, and whose mathematical formulation is very similar to the equation postulated by Newton in order to describe the gravitational attraction between two bodies. Further, in the field of heat conduction Fourier made significant progress by applying mathematical techniques to the problem of heat flow in different materials. The fact that Fourier considered the flow of heat is interesting and illuminating, since the concept of "flow" is taken straight from mechanics; the change in heat distribution over time was considered in analogy with a liquid, which through its flow becomes evenly distributed in space over time. This analogy was later carried over to the concept of an electrical current, i.e. the flow of electrons through a conducting material. Both these descriptions of physical phenomena highlight the importance of analogies originating in classical mechanics when laws and later theories in thermodynamics and electromagnetism were discovered.

Possibly the most obvious example of the transfer of ideas and concepts from classical mechanics is Maxwell's derivation of the equations describing an electromagnetical field. In the modern exposition of electromagnetical field theory, electricity and magnetism are considered to be vector fields in space,[5] whose direction and magnitude depend on a set of so called partial differential equations. In teaching, these equations are presented without further motivation, and the focus lies instead on solving them in particular cases. But the equations weren't simply postulated by Maxwell, but in fact derived from a mechanical perspective, in which he considered the magnetic field as a fluid full of vortices whose rotation was proportional to the magnitude of the field. The fluid was assumed to be electrically charged and the fluid

---

[5] A vector field describes a quantity which at every point in space has both a magnitude and a direction, as opposed to a scalar quantity which only has a magnitude. For example *wind* is a vector field, while *temperature* is a scalar field.

flow which derived from the vortices gave rise to an electric current. Based on this analogy Maxwell could with the help of the laws of classical mechanics derive the properties of the magnetic field. The fact that the magnetic field in reality consisted of something other (exactly what was unclear) than an electrically charged fluid was beyond doubt, which can be seen from the following passage taken from Maxwell's seminal paper *On Faraday's Lines of Force* (1861):

> The substance here treated of must not be assumed to possess any of the properties of ordinary fluids except those of freedom of motion and resistance to compression. It is not even a hypothetical fluid which is introduced to explain the phenomena. It is merely a collection of imaginary properties which may be employed for establishing certain theorems in pure mathematics in a way intelligible to many minds[...]

This clearly illustrates how useful analogies and models can be in order to grasp and think about natural phenomena, which otherwise seem intangible. It is also worth mentioning that the modern exposition of electromagnetical field theory in terms of partial differential equations was formulated in the 1890s by Hertz, who incidentally was the first to verify the theory experimentally.[6]

Maxwell also contributed to the development of a completely different field in physics, the kinetic theory of gases, and also in this case with the aid of mechanical analogies. That gases consist of particles, and that the temperature of a gas depends on the velocity of the constituent particles was suggested by Clausius, who also identified the temperature as the mean velocity of the particles. By viewing the particles as inelastic bodies that, in analogy with billiard balls, exchange momentum upon collision, Maxwell was able to derive their distribution of velocities. This result was later refined by Boltzmann, and the distribution is today known as the Maxwell–Boltzmann distribution.

The general approach that the above examples highlight was to first represent a phenomenon in a mechanical fashion and then, based on that arrangement, derive a mathematical expression that links the variables (e.g. pressure and volume) that describe the system. The importance attributed to this method can be appreciated by considering the following quote from 1904 by Lord Kelvin, one of the greatest physicists of the 19th century:[7]

> It seems to me that the test of 'Do we or do we not understand a particular point in physics?' is 'Can we make a mechanical model of it?'

It is worth pointing out that the physicists of the 18th and 19th centuries did not believe that the world was made up only of fluids, pulleys and billiard balls, but that it could be described as if it was. Mechanical models and analogies were a means of representing and visualising relationships between different parts of physical systems, and this in conjunction with classical mechanics made it possible to derive mathematical descriptions of these systems.

---

[6]H. Hertz, *Untersuchungen über die Ausbreitung der elektrischen Kraft* (1893).

[7]Baltimore lectures on wave theory and molecular dynamics (1904). In: Kargon, R. and Achinstein, P. (ed.) (1987). Kelvin's Baltimore Lectures and Modern Theoretical Physics, MIT Press (1987).

Abstraction

During the early 20th century the mechanistic school of thought had to yield to a more abstract and mathematical approach to physics. In this period quantum mechanics, which describes the physics of the very small, emerged as a contender to classical Newtonian physics, which many believed could explain all aspects of the world and was close to being a complete theory. The emergence of quantum mechanics effectively up rooted that belief and also turned out to be essentially different from classical mechanics. The predictions made by this new theory were not solid and deterministic statements, but only given in terms of probabilities. This indeterminism left little room for understanding the theory in terms of everyday mechanical analogies, and the contemporary relativity theory with its four-dimensional curved space-time was also difficult to grasp from an intuitive mechanical perspective. This displacement from the mechanical to the abstract and mathematical was summed up in 1937 by the philosopher of science Philipp Frank when he wrote:[8] "The world is no longer a machine but a mathematical formula".

In the work of many physicists emphasis was slowly shifting from analogies to theories and abstract models, and the former were considered to be lower rank, half-done theories that were only useful as tools for reaching a complete theory. A contributing cause to this was the unity of science movement within philosophy of science that advocated a rigorous unification of all the sciences. In this grand plan there was no room for small models of isolated phenomena, even less so for several distinct models of the same phenomena. Discovery driven by analogies, also known as the intuitive-transductive method, was challenged by the hypothetical-deductive method, inspired by contemporary trends in mathematics that strived for axiomatisation and formalisation.[9] The basic tenet of the latter method was to mathematically deduce experimentally testable predictions from a small number of assumptions or axioms about the phenomenon. This was however a highly set goal that, except in a few cases such as Fresnel's deduction of optical patterns of diffraction, was seldom realised[10] and mostly existed in the minds of philosophers of science. The limits of our knowledge are in general advanced by hypotheses and educated guesswork, rarely through grand derivations or deductions.

---

[8]The mechanical versus the mathematical conception of nature, 1937, Philosophy of Science 4: 41–74.

[9]This trend was most clearly expressed by the French Bourbaki group who published a series of nine books on different aspects of mathematics in which everything was solidly founded on set theory. The books were all published under the pseudonym Nicolas Bourbaki.

[10]Fresnel, A. (1818). Le diffraction de la lumière.

Followers

The other branches of science soon continued in the footsteps of physics. Chemistry followed a similar mathematical and abstract route, while other disciplines adopted the mechanistic thinking of physics without being able to adapt and apply mathematics to the same extent. A general trend that can be traced in the history of science is a transition from an aspiration to describe *how* the world is constituted to a desire to explain *why* this is the case. We want to know both how the world is constituted and why this is so. An explicit, albeit heavily simplified, example of this is the transition that biology went through in the 19th century. Linnaeus, who introduced the notion of a formal classification into biology, catalogued and classified animals and plants into different species, families and other taxa. Since it was considered that all plants and animals were created by God and forever fixed, there was no impetus to explain why a certain species existed in a given location or why it had certain characteristics. All this changed when Darwin, during the second half of the 19th century, discovered that novel species are created and shaped by evolution through natural selection. Biology changed from a discipline concerned only with classification to one which tried to explain why the organic world is structured the way it is. Darwin established the basic mechanisms necessary for evolution to take place: heredity, variation and natural selection, and argued that these mechanisms were sufficient to explain all aspects of biology.

Another branch of science that underwent drastic changes during the 19th century was medicine, which went from a fairly non-empirical discipline, based on vague, often untested assumptions about the human body, such as humorism,[11] to a quantitative discipline that strived to established empirical well-founded causes. An example of this is the discovery, made during the second part of the 19th century, that it is microbes, such as bacteria and viruses, that cause contagious diseases. The microbes thus were the mechanism by which diseases could be transferred from infected to healthy people, and with this knowledge it became easier to explain how diseases spread and also to contain outbreaks. Mathematical approaches have also been successful in medicine, and one example of this, related to contagious diseases, is the application of mathematical modelling to the spread of cholera in Soho during the Victorian era.[12]

After these two specific examples let us take a step back and think about the way in which explanations of almost all natural phenomena have changed over time. During the course of history, humans have always put forward explanations as to why the world looks the way it does. Initially these explanations were of a religious nature, but as time has passed scientific explanations have become increasingly common. But where in this chronology do models fit? When did models become commonplace in scientific explanations? A comprehensive answer to these questions requires a historical study that is far beyond the scope of this book. However, the latter question

---

[11] This theory states that the human body consists of four different substances: blood, phlegm, yellow bile and black bile, and that imbalances in these cause disease.

[12] Körner, T. (1996). The Pleasures of Counting, Cambridge University Press.

hints at a transition between different attitudes towards reality. How do simple models and simple explanations, such as humorism, really differ? An explanation that is based on a model is a *simplification*, which, despite the fact that it neglects many aspects of reality, provides a satisfactory answer, while a simple explanation assumes that the world in itself is simple. Humorism postulated that the human body only consisted of four distinct elements, not that the body acted as if it was that simple. Theories, be they simple or complicated, have a claim to truth, while models are constructed out of simplified and false premises. In other words, the two types of explanations differ because there is a fundamental difference between claiming that something is simple, and claiming that it can be explained in a simple way. Many complex systems are governed by surprisingly simple principles, and these are made use of when building models of these systems. Creating a model and using it in order to explain a certain natural phenomenon can be viewed as conceding its complexity, and the emergence of models in scientific practice can therefore be said to coincide with this insight. The world is immensely complex, but at least we can understand small parts of it.

Mechanisms

Previously in this chapter we have discussed the idea of "the world as a machine" and applications of this idea in terms of mechanistic explanations of natural phenomena. This method was first applied in physics (or rather classical mechanics) and has from there spread to other branches of science, and today it is difficult to imagine any kind of science that does not provide its explanations in terms of mechanisms. But what do we actually mean by the concept "mechanism"?

The exact meaning of the concept is difficult to pinpoint, the main reason being that it is used by so many disciplines in many different contexts, but a common feature is that it is used when one tries to explain an often complicated phenomenon in terms of simpler steps of interacting components. The crucial thing is that the system that exhibits or gives rise to the phenomenon consists of different parts and that these parts influence each other in a systematic and predictable manner. The mechanism is realised by a collection of objects, each with its own distinct properties, that interact with one another through a set of activities, and these interactions change the properties of the objects in such a way that the phenomenon comes about.[13]

In order to clarify this somewhat abstract reasoning we shall present a few mechanistic examples from two distinct disciplines, molecular biology and economics.

In all living cells proteins are being produced, and the structure of each protein, and therefore its function, is determined by a particular stretch of the genetic code, the DNA. The protein is however not produced directly from the DNA sequence but via a number of intermediary steps: initially the DNA is transcribed into messenger

---

[13]Machamer, P., Darden, L. and Craver, C.F. (2000). Thinking about mechanisms. Philosophy of Science 67:1.

RNA that is transported to the ribosomes outside the nucleus, where it is translated into a chain of linked amino acids that finally folds into the protein. This is a severe simplification of the actual sequence of events, which is highly complex and affected by a large number of additional factors, but at least this description gives an indication as to what a mechanistic explanation in molecular biology might look like.

The relation between interest, inflation and unemployment rate is commonly explained in the following terms: If the unemployment rate is lowered in a society then the total purchasing power is increased. This implies that the value of money is reduced and leads to an increase in inflation. On the other hand, the interest rate influences both inflation (higher interest rate implies less borrowed money, which increases the value of money and reduces inflation) and unemployment rate (lower interest rate leads to fewer investments and hence fewer jobs).

What we take as a mechanism depends on what we view as fundamental in our description of a system. Initially mechanisms were exactly what you would expect from the name, since mechanical activities (forces, torques etc.) and macroscopic objects, which could be idealised into levers or inclined planes, were the only ones known to the scientists of the 18th century. As more objects (cells, molecules, atoms) and activities (diffusion, exchange of electrons, enzymatic activity) have been discovered these have been added to the list of possible mechanisms. What we view as fundamental often depends on the scientific discipline we belong to. A molecular biologist can for example say the following concerning the function of a molecule:[14]

> Inositol triphosphate is a hydrophilic molecule and moves in the cytoplasm. This messenger works by mobilizing calcium from calcium stores in the endoplasmic reticulum. It does so by binding to a receptor and opening up a calcium ion channel. Once the ion channel is open, calcium ions flood the cell and activate calcium dependent protein kinases...

In this case we can identify molecules binding, opening and cleaving as the mechanisms at work. Chemists usually make use of more fundamental or lower level mechanisms, as in the following example that describes how an alcohol is dehydrated in the presence of an acid and a water molecule is produced:[15]

> The reaction begins with a Lewis acid–Lewis base reaction between a hydrogen ion from sulfuric acid and a nonbonding electron pair on the alcohol: an oxonium ion results. The oxonium ion loses a water molecule to form a carbocation. In the final step, the carbocation is neutralized by elimination of a hydrogen ion with the resultant formation of a carbon–carbon double bond.

In this example the mechanisms act at the level of single electrons, atoms and molecules, and result in the neutralisation and creation of chemical bonds.

These four examples from molecular biology, chemistry and economics illustrate how the behaviour of a system can be broken down into interactions between subsystems that together give rise to the dynamics of the whole system. To a certain degree these explanations can be viewed as models in themselves, since they

---

[14]Patrick G.L. (2001). An introduction to medicinal chemistry, Oxford University Press.

[15]Bailey, P.S. and Bailey, C.A. (2000). Organic Chemistry: A brief survey of concepts and applications, Prentice Hall.

give a simplified and idealised picture of the true course of events, but importantly this reduction of the phenomenon implies that it in a natural way can be translated into mathematical terms. A system whose behaviour is driven by mechanisms can be modelled, whether the mechanisms are of a "classical mechanics" type (billiard balls etc.) or of more recent kind (chemical reactions, diffusion etc.). The dynamics of the system can be described either verbally (as above), graphically as a flow chart or quantitatively with one or more mathematical equations.

Christopher Polhem—A Modelling Pioneer

We will now take a leap back in history and focus on one of the pioneers of Swedish science and, in particular, modelling.[16] Christopher Polhem was active during the first half of the 18th century. He worked as an engineer, inventor and scientist, and is considered one of the most prolific scientists to have ever lived. Polhem was first and foremost interested in practical things, such as identifying and solving technical problems in mining, transport and handicrafts. These tasks suited him very well, since he was equipped with a well-developed sense of space, and had an intuition for mechanics that could be matched by none. In his youth, with the support of a stipend from the Swedish Board of Mines, he travelled around Europe for five years. The purpose of the trip was both to educate the young Swede, and, in a form of industrial espionage, to acquire knowledge about technical novelties and innovations, such as mills, saw mills, and locks. Without making a single drawing during his trip he could, upon his return to Sweden, reconstruct several machines simply from memory.

Although Polhem had an excellent intuition for mechanical construction he was not alien to the idea of systematically investigating his inventions to see if there was any room for improvements. In order to achieve this he made use of miniature scale models that he collected at his *Laboratorium mechanicum* in Stockholm. It was a place intended as a source of inspiration for other engineers and inventors, but also had an educational purpose, where students could learn basic mechanics in a concrete manner. The most famous of Polhem's model constructions is the "hydrodynamical experimental machine" that was built in order to investigate the efficiency of different water wheels. By letting water flow through the wheel for exactly one minute, and at the same time measuring the velocity of the wheel, the efficiency of the design could be estimated. The construction contained five parameters that could be adjusted independently: the shape of the shovels, the ratio between the diameter of the water wheel and the crank, the drop of the water, the angle of the water inlet and finally the load put on the wheel. By systematically varying these parameters in repeated experiments (it has been estimated that 20,000–30,000 experiments were carried out during two years) Polhem tried to find the parameter setting that yielded maximal power (energy per unit time) as a function of the wheel load. This is most likely

---

[16]David Dunér (2009). Modeller av verkligheten. Modellbyggaren Polhem, seendet och det spatiala tänkandet, Vetenskapssocieteten i Lund, Årsbok.

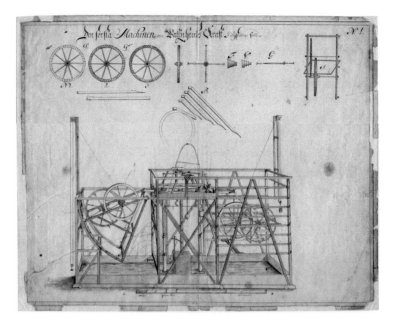

Fig. 1 A drawing of Christopher Polhem's "Hydrodynamic experimental machine". *Source* The library of the Royal Institute of Technology, KTH, Stockholm

the first documented case of optimisation through parameter variation, a technique that today is common, and is often used in the context of mathematical models and computer simulations.

A curious story related to the hydrodynamical experimental machine concerns a former student of Polhem's, Per Elvius Jr., who also tried to optimise the construction, but this time with the help of classical mechanics. With the aid of Newton's laws of motion and mathematical deduction, Elvius tried to find the most efficient construction without actually carrying out a single experiment. Interestingly he was strongly discouraged to try this by Polhem himself—a distinct clash between practical and mathematical modelling.

Polhem was also the first engineer to suggest that experiments with scale models of ships in a flume could be used in order to investigate how the shape of the hull affects the water resistance of a ship. We know today that he was well aware of the problems that appear when scale models are utilised,[17] but most likely he was ignorant of the complicated scaling that one needs to take into account when modelling the flow of

---

[17]The problem of scaling is best illustrated by looking at how the surface area and mass of a model changes as the size of the model increases, a relation which is of major importance when building e.g. prototypes of airplanes. The volume and therefore mass will increase as the length to the third power (if the length is doubled the mass increases eight-fold) while the surface area of the wings and hence the force lifting the plane only increases as the square of the length (and only quadruples when the length is doubled).

water (see p. 89). He did however not conduct any quantitative studies, but instead focused on the qualitative impact of the design of the hull on the properties of the ship. For example he tried "either a pointed, broad, flat, hyperbolic or parabolic hull and a high, low, pointed or broad bow". The actual experiments were not carried out by Polhem, but by his assistant Emanuel Swedenborg, who today is at least as well known as his teacher.

The use of scale models, to let the large be represented by the small, and the idea that the same laws apply to both is something that we today take for granted, but this was far from obvious for the contemporaries of Polhem. His achievements in engineering and his use of scale models makes him a pioneer of scientific modelling and his research paved the way for the modern use of models (Fig. 1).

Structure, Relation and Use

In the previous chapter we looked at the historical roots of scientific models. We will now turn to their contemporary use and view them from a philosophy of science perspective. As a starting point for this discussion we will specify what we actually mean by models, and discuss the different classes of models that exist. In order to broaden the concept and connect it to other concepts within science, we will relate modelling to phenomena and theories, and also analyse the relation between simulations of models and experiments. These notions will all be tied together, and made concrete in a worked example, where we show how a model typically is constructed. Lastly, we will try to draw conclusions about desirable properties of models, and discuss the dangers of modelling.

Since this chapter focuses on discussing general concepts that both span different types of models and their use in many different disciplines, the contents lean towards the abstract. To discuss and analyse the general ideas we have had to sacrifice the particulars. Still it is our hope that the examples contained in the chapter will illuminate the problems discussed and clarify the conclusions drawn. Concrete examples of different types of models are also presented in the chapter Worked Examples.

The Concept of a Model

In order to discuss and reason about scientific models on a deeper level we initially need to define what we actually mean by a model.

This has turned out to be more difficult than one would imagine, the main reason being simply the breadth of models that exist. What does a scale model of a ship have in common with a flow chart illustrating a geological phenomena? In some ways a model represents a subset of reality with a certain end goal in mind, although the manner in which it does so and its accuracy might vary greatly. Let us briefly mention a few definitions that previously have been put forward:

© Springer International Publishing Switzerland 2016
P. Gerlee and T. Lundh, *Scientific Models*, DOI 10.1007/978-3-319-27081-4_3

> A model is an interpretative description of a phenomenon that provides access to the phenomenon.[1]

> ...[Models are] physical or mental systems that reflect the essential properties of the phenomenon being studied.[2]

> ...a common notion is to view scientific models as representations of separate real-world phenomena.[3]

These suggestions point in the same direction, and we will use the following definition, which is similar to all of the above, namely: *Models are descriptions, abstract or material, that reflect or represent, and hence provide access to, selected parts of reality.*

We have added the clause "and hence provide access to" to exclude trivial representations that do not provide any understanding of reality. The word "bear" can be said to represent a subset of reality, namely all animals known as bears, but the word itself does not give any insight into the nature of bears nor does it provide us with any access to the properties of the animal.[4] On the other side of the spectrum we can imagine extremely complex, but completely pointless, representations that do not provide any meaningful access to the phenomena under study. In fact this reasoning, if taken to its logical extreme, implies that anything can be a model of anything else. For example we could let a tea kettle be a model of an elephant (they both have mass and spatial extension, and might even have the same colour), but this would hardly give us any insight into the nature of an elephant.

At this point, it might be a good idea to stop and think about what is actually meant by the phrase "selected parts of reality". We take it to denote a phenomenon, a concept which in itself is not unproblematic, and will be discussed at length later in this chapter.

Model Taxonomy

Models can be divided into separate classes depending on how they represent the phenomenon in question. A common division is into conceptual, iconic, analogous, symbolic, phenomenological and statistical models.

Conceptual models are the most basic type and serve as the foundation for more concrete and mathematical models. They manifest themselves as ideas and notions about mechanisms and entities within a system, sometimes purely mental constructs, but often expressed in everyday language. For example the greenhouse effect is

---

[1] Bailer-Jones, Daniela (2009). Scientific Models in Philosophy of Science, University of Pittsburgh Press.

[2] Hansson, Sven Ove (2007). Konsten att vara Vetenskaplig.

[3] Sellerstedt, Bo (1998). Modeller som metaforer, Studier i Kostnadsintäktsanalys, EFI, Stockholm.

[4] The exception here is possibly onomatopoeic words, whose sound imitate their meaning, e.g. 'crow'.

often expressed in linguistic terms as "the gases in the atmosphere of a planet block the outward radiation from the surface of the planet and this leads to an increased average temperature". This conceptual formulation might be useful if one wants to communicate the model, but it can also be put in mathematical terms that allow for predictions with higher accuracy. With a mathematical formulation of the greenhouse effect it is e.g. possible to calculate how many degrees the average temperature of the Earth would increase by if the concentration of carbon dioxide was doubled. Another example of a conceptual model is the description of how stars are shaped by a balance between gravitational forces that pull matter towards the centre of gravity, and radiation pressure that pushes matter outwards, which (in most cases) leads to a stable equilibrium. This is also a general formulation that can be made quantitative by dressing it in a mathematical guise.

Iconic models are often viewed as the simplest type of models since they are direct representations of a system, only magnified, miniaturised or projected. To this class belong scale models, pictorial models and blueprints. The main reasons why iconic models are used is that they are far easier to handle than the original systems, and they save both time and money. Instead of building a full-scale version of the hull of a ship, and testing its properties, one is often content with constructing a scale model with which similar tests can be carried out, and based on the outcome, conclusions about the properties of the real ship can be drawn. In some cases the model is scaled in time and not space, such as the case of experiments with bacterial populations, with a short generation time (a couple of hours), which makes it possible to investigate phenomena on evolutionary time scales. Iconic models also allow for visualisation of certain systems, both in terms of miniatures, such as blueprints, and magnifications, such as atoms being modelled as plastic spheres that can illustrate how molecules are structured and interact.

Analogous models are not classified by their structure, but rather by the process of construction. The common feature is that they have been devised in analogy with a known system. Some striking examples of this are the use of hydraulic models to describe the flow of money and value between actors in economic systems, models of atoms that describe the orbits of electrons in analogy with the planets orbiting the sun, and the perennial harmonic oscillator. The latter model has been used to describe a wide array of physical systems, such as the dynamics of a simple pendulum or how molecules absorb electromagnetic radiation by storing the energy as internal oscillations (the phenomenon on which microwave ovens are based). The essence of an analogous model is that the system one tries to describe is compared with another system whose dynamics are known, and this knowledge is then transferred to the new system, which hopefully leads to novel insight. These models spring directly from the method of modelling that arose among physicists during the 19th century, where classical mechanics was used as a means to construct useful analogies. An analogous model can be both mathematical and informal. For example certain properties of light can be explained by making analogies with undulations on the surface of a fluid or in a gas, but these explanations will never be exhaustive since the two phenomena (light and ripples) are fundamentally different.

Symbolic models make use of symbols and formal systems in order to describe phenomena. A typical example is a mathematical model that utilises equations in order to represent the properties of a system. Such equations can vary both in number and kind, but the basic idea is that they describe features and components of a phenomenon that are considered interesting, and interactions between these variables. There are models that only consist of a single equation, such as the Fisher model, which describes how an advantageous genotype spreads in a population as a function of time and space, but also models that are composed of hundreds of equations, such as models of genetic regulation in molecular biology. A common type of equations, especially within physics, are partial differential equations, which describe how continuous quantities such as a magnetic field or wind speed changes in space and time. Also discrete models in the form of difference equations are common. In addition to models formulated in terms of equations, there are also agent-based models, which have become popular within economics and biology. These models are defined by assigning the agents properties and rules according to which they act under certain conditions. This makes them difficult to understand from a mathematical perspective, and instead simulations are required to analyse their behaviour.

Phenomenological models are often symbolic in nature, and the two types of models are differentiated by their structural similarity. With symbolic models one tries to capture interactions between relevant, and in reality existing variables, while phenomenological models are used when the end result is prioritised and capturing the actual mechanisms within the system is viewed as less important. Such models, where only the outcome is of importance, are often viewed as "black boxes", since the internal mechanisms are considered uninteresting and hence can be neglected. An example of a phenomenological model is the Gompertz growth model, which is used for describing how the size of a growing population changes over time. It was originally derived as a means to estimate the expected life span of humans, and is based on the assumption that "…the average exhaustions of a man's power to avoid death are such that at the end of equal infinitely small intervals of time, he loses equal portions of his remaining power to oppose destruction".[5] Its most common use is however to model the growth of populations, and in particular the growth of tumours. Another phenomenological model that has been applied in cognitive science is the concept of the brain as a computer. The structural similarity between a brain and a computer is non-existent, since the brain consists of billions of neurons that process information in parallel, while in a computer information is processed by a central processing unit. Despite this, certain similarities can be seen between the composite behaviour of all neurons and the computations that take place in a computer.

Statistical models are a subset of symbolic models that make use of tools from probability theory. A statistical model of a system describes relations between variables that are not fixed but take on values from a certain distribution. These distributions are characterised by a number of parameters and statistical modelling often

---

[5]Gompertz, Benjamin (1825). "On the Nature of the Function Expressive of the Law of Human Mortality, and on a New Mode of Determining the Value of Life Contingencies". Philosophical Transactions of the Royal Society of London 115: 513–585.

boils down to estimating such parameters. With the aid of statistical models it is possible, by observing and analysing data from a phenomenon, to determine which interactions are important and what variables are relevant.

Let us briefly return to the example about rain and umbrellas that was discussed in the introduction (see p. 6). If we were to describe the situation with a statistical model we would proceed by introducing a stochastic variable X that represents the amount of precipitation that will fall during the entire day. This stochastic or random variable has a certain distribution, which is a function that assigns a probability to each level of precipitation. Initially we need to estimate this distribution by for example consulting previous meterological data. In order to avoid the rain that might

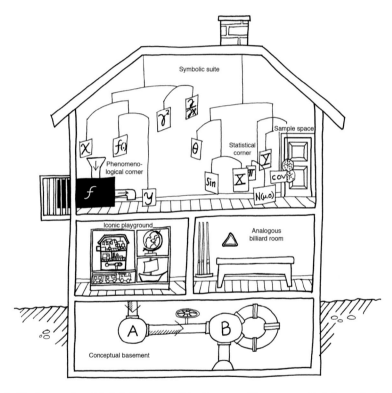

Fig. 1 The house of models is built on top of the Conceptual basement, which forms the basis for quantitative modelling. The ground floor consists of two rooms. On the *left* is the Iconic playground where one can study and play with scale models of the Earth, ships, trains and molecules. On the *right* is the analogous billiard room where the effects of every single shot can be predicted with perfect precision, the winner of the game is even announced in advance. The first floor has a bright and open design and fancy mobiles hang from the ceiling of the Symbolic suite. The *left corner* is considerably murkier due to the big *black box* and the functionalist design. On the *right side* in the Symbolic suite we find the Statistical corner with direct access to the Sample space. For more information about the house of models we would recommend a visit to the Iconic playground where one can study a detailed scale model of the house itself. © Lotta Bruhn based on an idea by Torbjörn Lundh

fall one could take an umbrella, but there is always the chance of leaving it behind somewhere. We let the probability that this occurs be represented by a variable Y, whose distribution depends on the degree of forgetfulness. When we are to leave the house we have to make a decision as to whether we are willing to take the bus (and pay the bus fare) or use the umbrella with the risk of losing it, were it to rain sometime during the day. If we know the distributions of the variables X and Y, the bus fare and the price of an umbrella, then with the aid of probability theory we can calculate the expected cost and choose the cheapest option.

This classification of models is hardly exhaustive and many models fall into several categories. This subdivision does however provide us with a framework with which to understand the similarities and differences between various models, and also how they relate to the systems they describe.[6] An artistic representation of the various classes of models that shows how they all fit under one roof is shown in Fig. 1.

Phenomena, Theories and Models

Up to now we have treated models as separate entities without any direct connection to theories, and with a straightforward relation to phenomena and related data. We will in this section try provide the reader with a deeper understanding of how the construction of a model relates to existing theories and the phenomena it is aiming to describe.

Grand Theories and Minor Models

Physicists have for a long time promised a "Grand Unified Theory" that will unite all forces of nature, and as a consequence all existing physical theories will become special cases of this one fundamental law of nature. The aim has been to move away from the specific towards the general, to understand that all forces and physical phenomena in fact are different sides of a single many-faceted coin. Implicit in this ambition is the sentiment that a theory that can explain as much as possible of the natural world is to be preferred. But does this notion extend to scientific models: do we hold the same preferences when it comes to their properties?

As the name suggests, models are small measuring devices, while theories are grand statements about the constitution of the world. In order to act as "measuring devices" that tell us something about a specific part of reality, models have to be carefully aligned with their purpose. An illuminating comparison could be made with a map. A single geographic area is usually covered by a collection of different

---

[6]Other classifications are also possible, e.g. into material, mathematical and fictitious models, as suggested by Gabriele Contessa in the paper *Scientific models and fictional objects* Synthese 172 (2010).

maps, each drawn with a particular purpose in mind. There are orienteering maps, for those who wish to run about off-road, bike maps for those who prefer to cycle and road maps for those who drive. These maps differ both in their scale and degree of detail. For the person running it is desirable to avoid hills and mountains since running around them saves time, but for a driver this matters less. Because of this, level curves are always drawn on maps for orienteering, but never on road maps. This reasoning also extends to models: there is always an underlying purpose of a model, and a certain aspect of a phenomenon is therefore being accentuated. What might happen when a model is constructed to incorporate all aspects of a phenomenon is vividly described in one of Louis Borges's stories.[7] In the story a group of cartographers are given the task of drawing a map of a fiefdom. After much work and discussion they finally agree that the most detailed and complete map was one on a 1:1 scale, such that when it was unfolded it covered the entire fiefdom, and hence was utterly useless as a map.[8] The moral of this story also applies to models. They have to be simple (at least simpler than the system they represent) and idealised in order to be useful.

Theories and Models

In order to get a deeper understanding of what models are and how they are employed in scientific practice, it is necessary to relate them to theories. Although models play a central role in science, theories are equally important.

Looking back in history it is in fact scientific theories that have dominated the thoughts and ideas of philosophers of science. How well this dominance reflects the workings of the individual scientists is not clear, but today it seems as if the actual scientific method consists more of "trial and error" rather than careful formulating of grand theories. The elevation of theories as an ideal has its roots in the positivism of the French 19th-century philosopher August Comte, and was later popularised by the logical positivist movement that developed during the 1920s in Vienna, which was spearheaded by Rudolf Carnap and Otto Neurath.[9] From this philosophical movement sprang the idea of "unity of science" that proposed to unify all strands of science into one complete theory. From the laws and theories of physics one would deduce the laws of chemistry, that would in turn lead to biology, and so on all the way to human psychology and sociology. The gaps between the sciences would be filled, and left standing would be one great theory that could explain all aspects of reality, from neutrons to neurons and nervousness. In this, somewhat pompous plan, there

---

[7]"Del rigor en la ciencia" in the collection of short stories *Historia Universal de la Infamia* (1946).
[8]This story is possibly inspired by Lewis Carroll who in *Sylvie and Bruno Concluded* (1893), Chap. 11, describes a similar situation.
[9]The name of the movement derives from the French word *positif* which roughly means "imposed on the mind by experience".

was no room for sloppy models with limited claims of explanation and applicability.[10] The development took a turn in the 1950s, after important contributions by Jerry Fodor among others, that highlighted how unreasonable the claim of unity in fact was, and advocated a more benevolent view of the special sciences.[11] The idea of unity is however still alive today and was recently popularised by the biologist E.O. Wilson in his book *Consilience: The Unity of Knowledge* (1998). But what is the position of theories today, and how do they relate to models?

Philosophers of science view a scientific theory as a collection of statements in the form of mathematical equations or sentences in natural language, from which one can, with a given set of rules, derive new statements that are always true in the system that the theory describes. The basic statements, or axioms, are expected to be as simple and few as possible. There shouldn't exist any other set of statements that are fewer and simpler out of which the current set of axioms can be deduced. How are these axioms to be chosen? One approach is to view the axioms as hypotheses, from which predictions can be deduced, that are then tested experimentally. This is known as the hypothetico-deductive method and has been successfully applied in classical physics. This strict view of theory fits physics, the favorite discipline of philosophers of science,[12] fairly well, but doesn't agree with how theory is used in other disciplines. A more general account of a theory states that it is a collection of concepts and observable phenomena together with rules or laws that describe how observations of the phenomena are related to the concepts. It is from these basic concepts that, at least non-phenomenological, models are built. The theory provides a framework with which models of specific phenomena and systems can be constructed. For example, if we want to construct a model for the mechanical load in a football stadium[13] we have to make use of concepts such as force, torque, compressive stress and shear stress, which are provided by the theory of classical mechanics.

However, if no theory of the system under study exists, then modelling precedes theory. By constructing and analysing models it is possible to generate new hypotheses and test preliminary version of a theory. In this case the model becomes a deductive tool for investigating the phenomenon and serves as a complement to experiments.

The ambiguous relation between model and theory can make it difficult to distinguish them from one another. Sometimes they can be misleadingly similar, and what separates a theory from a model is sometimes down to the name given to them. The Standard Model in fundamental physics that describes the relations and interactions between elementary particles does not contain any simplifications and idealisations and is more akin to a theory, while theories in economics often contain substantial

---

[10]This is a condensed and simplified account of the dynamic debate about the concept of unity that took place within the logical positivist movement, see e.g. the debate between Neurath and Kaller in a series of papers in the journal *Philosophy and Phenomenological Research* in 1946.

[11]Fodor, Jerry (1974). Special sciences and the disunity of science as a working hypothesis, Synthese 28.

[12]Philosophers have viewed physics as a model science in the archetypal sense of the word.

[13]Often with supporters jumping in synchronization.

simplifications and idealisations, and therefore ought to be denoted models. There is also a tendency to view models as half-baked theories. If the model was further developed, and properly experimentally verified, it would eventually turn into a theory. The motivation for this point of view is ambiguous. Certain models, such as the model of bacterial growth discussed in the introduction (see p. 2), can never grow into a theory (the "theory of bacterial growth"). It is too simple and limited in its scope, and to reach the generality expected from a theory it would have to be modified beyond recognition, and probably acquire such complexity that it would lose its initial purpose. The conclusion is then that only a subset of models can aspire to become theories, but in our view this is inconsistent, and only leads to a fragmentation of the concept. Models can be used in order to form theories (a viewpoint that is verified by the scientists we have interviewed), but do not themselves transform into theories.

At this stage we have a better view of what actually constitutes a theory, and what separates it from a model. It would therefore be illuminating to compare the features that are considered desirable for a theory and a model respectively. How ought a theory or a model to be constituted? For theories this is fairly easy to answer. They should be:

1. Correct: they provide a description that agrees with our observations of the phenomenon in question
2. Complete: they should describe all aspects of the phenomenon
3. General: they should be valid for a wide range of phenomena
4. Simple: they should consist of a small number of assumptions and be easy to work with.

In contrast to these demands we propose the following list of desirable features for models:

1. Correct or incorrect
2. Incomplete
3. Specific
4. Simple or complicated.

This somewhat strange wish-list requires some motivation. Naturally we would like to have models that provide an accurate description of reality, but this is not their only purpose. For example it can be useful to construct a model that is false, since it could provide knowledge as to what important parts of the phenomenon have been left out (there are several advantages of false models and these will be discussed on p. 44). Secondly, it is not expected that a model should supply a complete description of a system, rather it represents a chosen aspect of the system. Further, models with a smaller range of applicability are preferable. This is usually specified by stating that a model is "for" a specific system or phenomenon. The degree of specificity is obviously a sliding scale and depends on a number of factors, such as the complexity of the system and the desired accuracy of the model. Lastly, there are no set rules when it comes to the complexity of the model itself, i.e. how many different variables and processes it takes into account, but this varies from one model to the next. Generally speaking one can say that models in physics tend to be simpler and minimalist, while models in e.g. biology, environmental science and medicine tend to be more complicated. This is far from a rule, but to some degree reflects the mode of operation within the different disciplines. Physicists have the opportunity to perform controlled experiments in which many variables can be kept constant, and therefore need fewer variables in their models, while variables in e.g. biological systems often are more difficult to control, which leads to models of larger complexity.

Yet another dividing line can be identified between theories and models. Generally speaking theories tend to be more abstract, while models deal with and represent concrete situations. Nancy Cartwright, a philosopher of science, even claims that because of their abstract nature theories do not tell us anything about reality.[14] Theories can only be used for making statements that are idealised to a degree that makes them impossible to test experimentally, and this is where models enter as mediators between theories and reality. Within the model, the theory becomes concrete by interpreting it into something tangible.[15] The only way to test the validity of a theory is to verify (or rather falsify) predictions within the theory, and this can only be done through experiments. However, in order to interpret the outcome of experiments we need a model of the experimental setup: the theory is applied to the specific situation, and this is precisely what modelling often is about. The outcome is then compared with the predictions from the model, and based on this we make judgements about the validity of the theory. From this point of view models are situated in between theories and phenomena, and therefore are an indispensable ingredient in the scientific method.

The Relation Between Phenomena and Models

In order to get a proper understanding of what models are, it is not sufficient to account for their structure and relation to theories, but one also needs to have an appreciation of what they are models of, in other words what they help us understand or describe. From an intuitive point of view one can say that phenomena are events, facts and processes that are clearly separated, such that they can be discriminated unambiguously. The degree of abstraction a phenomenon exhibits can vary greatly; we include both concrete events such as the currents of water in a specific river delta, and highly abstract activities such as life and metabolism. An interesting detail in this context is that scientists rarely have direct access (via sensory data) to the phenomenon, and instead they have to rely on external data that describe the phenomenon one way or another. In addition to this the situation is made more complicated by the fact that what is being perceived as a phenomenon depends on our current view of

---

[14]Cartwright, Nancy (1997). Models: The blueprints for laws, Philosophy of Science, 64.

[15]Or in the words of Ludvig Wittgenstein in the *The blue and brown book* (1958): "It has been said that a model in one sense covers the pure theory; that the *naked* theory consists of statements and equations."

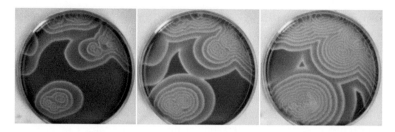

Fig. 2 A time series of a pattern formed by the Belousov–Zhabotinsky reaction. Published with kind permission of © Peter Ruoff

reality, and only phenomena that fit the prevailing perception of nature are viewed as real.[16]

An example of this is the so called Belousov–Zhabotinsky reaction that produces intricate spatial patterns from an initially homogeneous mix of chemicals (see Fig. 2). This phenomenon seemed to contradict the second law of thermodynamics, which states that in a closed system the degree of order always decreases with time. A homogeneous mix should according to this remain homogeneous for all future times. The phenomenon therefore didn't agree with the existing scientific view and the accounts of the patterns were heavily criticised when they were communicated by Belousov during the 1950s. It took another 10 years before Zhabotinsky took Belousov's initial observations seriously and could theoretically explain how they appear—and no, the second law of thermodynamics still holds.

Another example of how existing theories affect how we view the world, and what is viewed as a phenomenon, can be found in the discovery of the particle-wave duality of the electron. That electrons can behave not only as particles but also as waves was suggested by de Broglie in 1924 in his doctoral thesis. This prompted a group of scientists in New York (of which C.J. Davisson and G.P. Thomson in 1937 received the Nobel Prize in physics) to interpret their experimental data obtained from electron/metal collisions as a diffraction pattern, which can only arise if electrons appear as waves.[17] The patterns in the experimental data existed several years before they were interpreted as diffraction, but before de Broglie's theoretical advancement they were not even considered as a phenomenon.

A demand that is often put on models is that they should resemble the phenomena they represent structurally. This implies that different parts of the model should correspond to distinct parts of reality that are relevant for the phenomenon. In some cases an isomorphic, one-to-one, relationship between parts of the model and parts of reality is required, where each component and mechanism in reality has a counterpart in the model (and the other way around). In most cases this requirement is too strict and difficult to fulfil, since complete information about the mechanisms acting within a system is often not available. Another issue with an exact representation of the real

---

[16]Kuhn, T.S. (1962). The structure of scientific revolutions.

[17]An account of the discovery can be found in C.J. Davisson's Nobel lecture from 1937.

system is that the model tends to lose its meaning, since it becomes as complex as the phenomenon, much like the map in Borges's story.

For scale models structural similarity is relatively straightforward; the hull of the model ship should, to the best possible degree, conform to the shape of the hull of the real ship, but also other properties, such as the centre of gravity and density, which are important for the fluid mechanical properties of the model, should agree. For similar reasons the angles of the chemical bonds in a model molecule should agree with those observed in reality, but in the case of abstract or mathematical models structural similarity can be difficult to perceive and verify. What we chose to include in a model depends on our perception of how important different processes or subsystems are; this could be difficult to assess a priori, and is a skill that is acquired from years of experience.

The above discussion has related the structure of models to the phenomena they represent, but another important aspect is the relation between model and experimental data. Firstly, the model has to be able to recapitulate experimental data within a certain degree of accuracy. This process is usually called model validation and acts as a proof of warranty. The number of experimental situations that are taken into consideration and the degree of accuracy required in the validation depends on the type of model considered. Mathematical models are usually expected to exhibit higher accuracy compared with conceptual models that only provide qualitative statements about trends etc. After the model has been validated it can be used for making predictions about novel experimental (or real-world) situations and to a certain extent it might replace experiments. A typical example of this are meteorological models that are used mostly as a means to predict the weather. Certain models are not used in this quantitative and predictive sense, and their purpose is rather to explain how a given phenomenon comes about. They are usually called explanatory models, and can be used in order to prevent the appearance of a phenomenon, such as a disease, by providing insight into what parts or processes of the real system are causing the disease. Explanatory models are also common in the social sciences and are often used in order to find out why a certain course of events, such as the outbreak of a war or a revolution, took place. The border between predictive and explanatory models is often blurry, and models that exists in the no-man's-land can in this context be attributed with two roles: on the one hand they serve as containers of knowledge (by explaining how a phenomenon comes about), and on the other hand as tools that make use of this knowledge (to make predictions).

Since models are simplifications of reality there are always parts of a phenomenon that a model cannot account for, or in other words there are always features of the real phenomenon that are not reflected in a single model. But the reverse situation might also occur: the model has properties that don't exist in the real phenomenon. However, this aspect of modelling is rarely discussed. What we mean is that the model might have properties that don't agree with reality, but these are often neglected in favour of the agreement between model and phenomenon. An animal model of a disease has a large number of features that are specific to the species that the animal belongs to and not present in humans. In a similar way scale models of atoms and molecules have visual (atoms of oxygen are red and hydrogen white) and tactile (smooth surface) properties that real atoms obviously do not possess. Those that use the models are in most cases aware of these discrepancies, but there is always a chance that they affect the conclusions drawn from the model. This problem has been discussed by Mary Hesse,[18] who coined the terms positive and negative analogies between phenomenon and model. There is also a third class that constitutes neutral analogies, whose presence in the real world is uncertain. Most desirable is of course to construct a model that only contains positive analogies, but this might be difficult to achieve in practice.

Creating and Using Models

We have so far discussed what models are and how they relate to phenomena and theories. In this section we will try to put these pieces together in order to get an overview of how models are created and used. As part of this discussion we will also touch upon the relation between experiment and simulation, and the importance of model parameters.

The Art of Model Building

Describing how models are constructed is far from straightforward, mainly because there are so many different types of models (iconic, symbolic, conceptual etc.), but also because the process varies between different scientific fields. We will therefore try to depict the process of modelling both from a general and hence abstract perspective, and by using a concrete example from physics. Further examples of model building and their applications are given in the Chap. "Worked Examples", which covers all the different kinds of models discussed in the taxonomy (p. 21).

The cornerstone of the model-building process is a phenomenon, something in the world one aims to understand or predict, and a theory, which provides the basic laws for the domain to which the phenomenon belongs. Based on these two, a model is formulated, in a process where both theory and phenomenon play a role. The phenomenon controls which parts of the theory are applicable and what approximations are reasonable, while the theory defines what types of entities, properties and interactions the model can contain.

Once the model is constructed it is desirable to validate the model by verifying that it gives an accurate description of the phenomenon. In order to achieve this the model needs to be compared with either existing data or new data that is generated from experiments involving the phenomenon. The outcomes of an experiment are however rarely in a format that is directly comparable to the model; instead it is

---

[18]Hesse, Mary (1966). Models and Analogies in Science. In: Edwards, P. (ed.) The Encyclopedia of Philosophy. New York: Macmillan, pp. 354–359.

usually termed "raw data". Before a comparison with the model is possible the raw data has to be processed, for example by filtering or removing obvious errors. This results in data that is directly comparable to the model and allows for the validation to be carried out. The degree of agreement between the data and the model that is required in order to consider the model validated is highly variable and depends on the purpose and scope of the model. A validated model can be used as a means to predict future instances of phenomena, or to gain insight into the dynamics of phenomena and hence allow for manipulating them in a desirable way. Another possibility is that the validated model can be used as a means to challenge and alter the existing theory. The entire process is depicted in Fig. 3.

In order to make this somewhat abstract description understandable and tangible we will imagine ourselves located in a physics laboratory where the task at hand is to make a model of a pendulum. To this end we are equipped, as an underlying theory, with the laws of classical mechanics, which tell us how bodies react when subject to

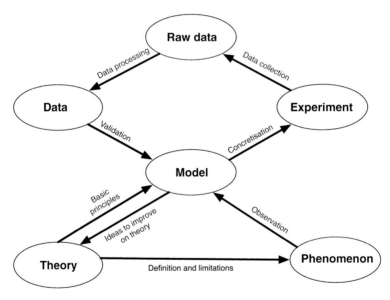

Fig. 3 A schematic description of the construction and use of models. Observation of reality based on a certain theory has led to the identification of a *phenomenon*. In order to create a *model* of this phenomenon basic principles are necessary, and these are provided by the underlying *theory*. In order to test the validity of the model it has to be compared with reality, and the situations in which the model applies are investigated with the help of *experiments*. In these experiments *raw data* is produced which after processing result in *data* which is directly comparable to the model. This process is known as validation, and if the model passes this test it is usually considered as satisfactory and can be used for making predictions or gaining insight into the dynamics of the phenomenon. If the model doesn't pass the test it needs to be modified; additional mechanisms might be required or the phenomenon needs to be restricted. It is also possible that the underlying theory isn't complete, and the model can in such cases provide ideas on how to improve on the theory

external forces. We construct the experimental setup and observe the phenomenon before our eyes—a weight suspended on a string oscillating back and forth.[19] What forces act on the weight at each moment in time? We make a simple drawing, in which we make a number of conscious and unconscious simplifications; the string is assumed to be without mass and completely stiff, and further we disregard the influence of friction of air on both string and weight. Drawing the setup in two dimensions we also make the tacit assumption that the movement only occurs within a plane. Once we have a clear view of the forces involved, and have separated them into tangential and vertical components, we can apply Newton's laws and write down the equations, which results in a mathematical model. It is now time for the experimental part, where by varying one property of the pendulum at a time we compare the experimental outcome with the predictions of the model. Let's say that we start by varying the length of the string, but keep the initial angle and mass of the weight constant, and measure the period (i.e. the time of one oscillation). This procedure generates a table of raw data, which can be plotted as a curve in a diagram, where the x-axis corresponds to the length of the string and the y-axis to the period. We are now in a position to compare the predictions of the model with the experimental data. Do they agree, or have we made a mathematical or experimental mistake somewhere along the way? And what do we even mean by "agree"? In order to answer these questions we have to analyse possible sources of error and decide to what extent we can expect the model and experiment to agree. Let us assume that we are content with the relation between period and string length that the model predicts and move on to the influence of the initial angle. We proceed as before and vary the initial angle, while keeping the other properties constant, and again plot the result as a diagram. What we see is that the period is essentially independent of the initial angle, at least for "small" angles. Why is this the case? If we return to the mathematical model we realise that this is because for small angles, the sine of an angle is roughly equal to the angle itself (when measured in radians).[20] This implies that for small angles we can simplify the model and reduce the number of parameters. This is however a rare event and often models have to be expanded in order to match reality. We are now equipped with a model, in terms of a mathematical equation, which describes how the period of a pendulum depends on the length of its string. This gives us a compact description of the phenomenon "pendulum" and we can now use this knowledge in order to construct a device which measures time, or apply the model in order to determine the strength of the gravitational acceleration. Interestingly, a simple, but very long pendulum (67 m in fact), was used by the physicist Léon Foucault in 1851 as means of illustrating the rotation of the Earth.[21]

---

[19]The pendulum is assumed to only oscillate back and forth, not to move in the third dimension.

[20]This is a first-order approximation in a Taylor series of the sine function, and corresponds to the approximation $\sin(x) \approx x$.

[21]Foucault, L. (1851). Demonstration physique du mouvement de rotation de la terre au moyen du pendule, Comptes Rendus de l'Acad. de Sciences de Paris, 32.

The above descriptions of model building are optimal in the sense that the researcher is assumed to have access to theory, phenomenon and model. But what happens to the procedure if any of these building blocks are missing?

In the case where no theory exists from which basic concepts can be drawn and the model built upon, the only thing that can be achieved is a phenomenological model of the data. Without any ideas about which mechanisms or processes govern the phenomenon at hand it is often not sensible to create a model that can do more than just predict the observables of the system, something that can still be quite useful. However, an alternative to the above approach is to devise a mechanistic model which might serve as a starting point of a theory for the system. The existence of hypothetical mechanisms and entities can be examined and tested in such a model. This approach might serve as a first step when a new phenomenon has been discovered which no existing theory applies.

There are also situations where the phenomenon under study is quite abstract, which might make it difficult to separate it from the model which represents the same phenomenon. This arises for example in the case of deterministic chaos which is usually studied with mathematical models, which although they originate from real-world phenomena such as weather systems or ecosystem dynamics, have been simplified to such a degree that they have lost all predictive power and only retain the "chaotic" features of the original system. Examples of such models are the Lorenz model[22] and the logistic map.[23]

Finally, there are cases in which the theory and phenomenon are in place, but a model simply isn't necessary. In these instances the phenomenon is of such a character that no simplifications of the theory are necessary in order to apply it and hence test it in an experimental setup. An example of this is the curvature of light from far away galaxies, due to the gravitational field of the sun, which was used as a means of testing Einstein's general theory of relativity.

Simulations and Experiments

Many mathematical models are so complicated that the only way to investigate their behaviour is by simulation. A typical example of this are the meteorological models that are used for weather forecasting. When a meteorologist on the evening news shows an animated sequence of how a cold front is closing in on your home town, the images shown are the result of simulating a meteorological model, where each image in the sequence corresponds to the state of the model at a certain iteration or time step. In a simulation the dynamics of the system are usually iterated forward in time, by making small steps in time, such that the state of the system in the

---

[22]Lorenz, E. (1995). The Essence of Chaos.

[23]May, R.M. (1976). Simple mathematical models with very complicated dynamics. Nature 261.

following time step only depends on the state in the current time step.[24] In the case of meteorological models the state is specified by the direction and strength of the wind, the air pressure, temperature etc. at a number of points distributed across the surface of the Earth. The calculations required to iterate the model dynamics were originally carried out by humans using only pen and paper, a procedure that could last for hours or even days. These days, with the aid of modern computers, the corresponding (or even more involved) calculations can be carried out within milliseconds.

One could even claim that computer-aided simulations have given us knowledge of the behaviour of models (and the phenomena they represent) that was previously inaccessible, and hence have revealed a whole new class of models that in earlier times were deemed unthinkable due to their complexity. Formulating complex models was of course always possible, but since it was impossible to investigate their behaviour this was considered a useless excercise. At least partially, the increase in the use of mathematical and symbolic models over the last few decades can be attributed to the increase of computational power. Despite these vast computational capabilities it is still the time it takes to simulate models that sets the boundaries. Climate models, which are typical examples of computationally heavy models, might for example require weeks in real time to reach the desired time point within the simulation. This is in stark contrast with models in the financial sector (where time is at premium) which have to be simulated within a second.

Simulations are in many respects similar to experiments, since they realise a given phenomenon and allow for its observation and characterisation. The drawback is of course that the simulation only allows us to view the idealised version of the phenomenon and not the real thing. On the positive side, simulations have many advantages: they are fast to carry out, it is easy to change parameter values, and they make it possible to study events that are impossible to carry out experimentally, such as the impact of asteroids and the dispersion of toxic agents in the environment.

In many cases it is desirable to formulate simple models that can be studied analytically, i.e. where it is possible to mathematically prove which properties the model does or does not have. Before the computer became popular in scientific work this was essentially the only method by which mathematical models could be studied. This approach has the advantage that it is possible to make definite statements about the properties of the model, without having to resort to time-consuming simulations, which have to be repeated as soon as the parameter values are altered. Validating simple models that can be treated analytically can however be difficult since in experiments many variables may be required to be held constant. An experiment where many variables are allowed to vary is most likely easier to carry out, but requires a more complicated model, which in turn is more difficult to analyse. Hence, there is an obvious trade-off between the complexity of the model and the experiment required to validate it.

---

[24]This is the case for so called deterministic models, but there are also stochastic models where randomness also determines the dynamics.

Parameters

An integral part in the construction and use of models is played by parameters, and this is particularly so in the case of quantitative mathematical models. Parameters take on numerical values and correspond to certain properties of the entities or the strength of interactions among entities in the model. In mechanical models they might be the mass or density of a body, or the viscosity of a liquid; in chemical models: the rates of reactions or diffusion constants of chemical compounds; and in biological models: the rate of growth of a population or the selective advantage of a particular gene. Mathematical models may contain anything from no parameters, as in "diffusion limited aggregation", a model of crystal growth,[25] to several hundred parameters as in models of gene regulation and cell signalling. As the number of parameters in a model increases, the accuracy of it decreases, since every single parameter is known only to a certain degree of precision. An important aspect of using models is therefore to investigate how sensitive the model is to perturbations of the parameter values, a procedure known as robustness or sensitivity analysis.

In rare instances the parameter values can be derived from simpler relations, e.g. the growth rate of a population can be related to the properties of single individuals, but in most cases the values need to be determined from experiments. In such a case yet another model is required, one that represents the experimental setup. Optimally this new model contains only one unknown parameter: the one to be measured. By comparing experimental data with simulation (or analytical) results of the model the expected numerical value of the parameter (with an error margin) can, with the help of statistical tools, be determined. In practice there are a plethora of different methods for parameter estimation, but it is important to understand that there is no model-free (or theory-free) way of estimating a parameter. Or in other words, it is not possible to measure something without any preconceptions of what it represents, since this "something" always originates from a model (or theory).

Another issue is caused by the fact that determining the parameters experimentally is both costly and time consuming, and many modellers therefore resort to estimations of the parameters based on similar parameter values in the literature or some type of basic reasoning based on prior knowledge of the system.

All Models Are False, but Is Anything at Fault?

The well-known phrase:

> All models are wrong, but some are useful.

---

[25]Witten, T.A. and Sander, L.M. (1981). Diffusion limited aggregation, a kinetic critical phenomenon, Phys. Rev. Lett., 1981, 47:1400–1403.

that was coined by the statistician George E.P. Box[26] is a good summary of the difficult question of the truthfulness of scientific models. All models contain simplifications and idealisations and are therefore strictly speaking false, but the way in which these simplifications are introduced makes the difference between a model that is useful and one that is not. To simply conclude that all models are false misses an important point, namely in what way they are false, and how this affects their usefulness. Below is a list, compiled by the philosopher of science William Wimsatt, that summarises all the different ways in which models can go wrong:[27]

1. A model may be of only very local applicability. This is a way of being false that occurs only if the model is more broadly applied to phenomena it was not intended to describe.
2. A model may be an idealisation whose conditions of applicability are never found in nature (e.g. point masses, the uses of continuous variables for population sizes, etc.), but which has a range of cases to which it may be more or less accurately applied as an approximation.
3. A model may be incomplete—leaving out one or more causally relevant variables.
4. The incompleteness of the model may lead to a misdescription of the interactions of the variables which are included, producing apparent interactions where there are none ("spurious" correlations), or apparent independence where there are interactions.
5. A model may give a totally wrong-headed picture of nature. Not only are the interactions wrong, but also a significant number of the entities and/or their properties do not exist.
6. A model may simply fail to describe or predict the data correctly. This involves just the basic recognition that it is false, and is consistent with any of the preceding states of affairs. But sometimes this may be all that is known.

These points ordered roughly in terms of increasing seriousness, and the first three can be said to be on a sliding scale since they, at least to some degree, apply to models that are considered useful, but if they are too grave they might render models useless. The remaining points list features that one would like to avoid altogether, mainly because they not only leave out relevant information, but also add details and results that are false.

The usefulness or applicability of a model should not be evaluated solely on the basis of how well it can make predictions, but also on what we can learn from it, in particular as a means to construct future improved models of the same phenomenon. Naturally there are cases when starting anew might be better than continuing with a completely inadequate model, but in most cases modelling is an iterative process. In this context an important feature of a model is its transparency, i.e. the degree to

[26]Box, George E.P. (1979). Robustness in the Strategy of Scientific Model Building. In: Launer R.L. and Wilkinson G.N. (ed.), Robustness in Statistics: Proceedings of a Workshop. New York: Academic Press, pp. 40.

[27]False models as a means to truer theories. In: Nitecki, M. and Hoffman, A. (ed.) (1987), Neutral Models in Biology. London: Oxford University Press, pp. 23–55.

which mistakes can be localised and corrected when the model needs to be improved. This can be compared to writing a recipe for a dish: one should write down the instructions and ingredients in a structured and intelligible manner, so that if the end result is undesirable one can easily go back and make the appropriate changes.

In this iterative process of modelling, what function can a false model serve? The following classification is also due to Wimsatt:

1. An oversimplified model may act as a starting point in a series of models of increasing complexity and realism.
2. A false model might suggest new mechanisms acting within the system and hence alternative lines for the explanation of the phenomenon.
3. A model that is incomplete may be used as a template for estimating the magnitude of parameters that are not included in the model.
4. An incorrect simpler model can be used as a reference standard to evaluate causal claims about the effects of variables left out of it but included in more complete models, or in different competing models to determine how these models fare if these variables are left out.
5. Two false models may be used to define the extremes of a continuum of cases in which the real case is presumed to lie, but for which the more realistic intermediate models are too complex to analyse or the information available is too incomplete to guide their construction or to determine a choice between them.[28]
6. A family of models of the same phenomenon, each of which makes various false assumptions, might reveal properties of the phenomenon that are robust and appear in some or all models, and therefore presumably are independent of different specific assumptions which vary across models. If a property appears in one model, but not in another, one may determine which assumptions or conditions a given result depends upon.

Based on this list we conclude that false models can be useful in a number of different ways, and that mistakes contained in models can serve as a source of inspiration for future construction of models. This throws light on another aspect of modelling that might seem problematic. It seems as if there is no simple and objective way of determining whether a model is useful, at least not in the way that a theory can be discarded if it produces false predictions. This problem seems to stem from the fact that models have a dual role in science: on the one hand they function as a means to represent knowledge about the world, and on the other hand they serve as tool to acquire this knowledge. Scientific objectivity can be applied to the knowledge we already have, but the means of acquiring this knowledge—our scientific strategies— always contain a certain measure of subjectivity. If we look at cell biology as an example, one might ask: should we use large-scale models that capture the dynamics of all the genes in an organism simultaneously, or should we construct smaller models of subsystems, that only after careful analysis and scrutiny are joined together into

---

[28] Here Wimsatt imagines the spectrum of models to be one-dimensional, which is an idealisation of the true space of possible models.

one great model? There is no objective answer to this question, and instead it is up to each scientist to decide what strategy works best.

The Right Thing or a Right Mess?

When constructing a model there is always the risk that it becomes so complicated that it loses its usefulness. But why, or more importantly, how does this happen? In order to answer this question we need think about the purpose of a model. Ultimately, models are used for understanding something, or in other words to acquire knowledge about a phenomenon. But understanding has an ambiguous meaning: it can be interpreted as knowledge about how the phenomenon will behave in the future, but also as knowledge about how the phenomenon comes about in the first place. Knowledge about a phenomenon then depends on both our ability to make predictions, and our understanding of mechanisms acting and giving rise to the phenomenon. In many cases the two are linked: a mechanistic explanation might enable prediction. But this is not always the case: it is possible to comprehend the basic physical laws that govern the throw of a die, without being able to predict the outcome. The reverse relation is also not necessarily true: it is possible to predict the adult height of a child, given the height of its parents, despite the fact that we are far from understanding the underlying physiological and genetic mechanisms that control the growth process. How then, does this discussion relate to models of phenomena?

Generally speaking a simple model that contains few components, equations etc. is easy to handle and understand, but it most likely produces poorer predictions compared with a more complicated model that might be more difficult to understand. Hence we can assume that the predictive power increases when the complexity of the model increases, and at the same time the comprehensibility is reduced when the complexity of the model increases. Since both predictive power and comprehensibility are necessary in order to have knowledge about the phenomenon, we can view knowledge as a product of these two properties of the model. This implies that most knowledge is reached with models of intermediate complexity (see Fig. 4b).[29]

This mode of reasoning can also be applied in order to classify different types of models. By placing models in a two-dimensional diagram where the axes represent the predictive power (x-axis) and comprehensibility (y-axis) they fall into different categories depending on their characteristics (see Fig. 5). Models that have a low comprehensibility and small predictive power are usually in a early stage of development (field I). Mechanistic models often have a high comprehensibility, but might differ in their ability to make good predictions (fields II and III). Lastly there are models that have a high predictive power, but a low comprehensibility (field IV), typically represented by phenomenological or statistical models. The latter are often

---

[29]This relates to the philosopher Daniel Dennett's concept of *real patterns*, that describes how we as humans find patterns in nature at the level on which they have the largest predictive power. Dennett, D. (1991) Real Patterns, The Journal of Philosophy, 88.

(a)   **(b)**

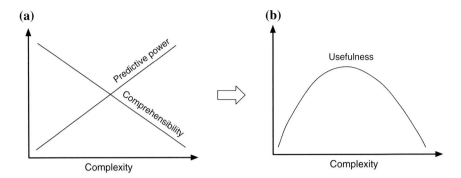

Fig. 4 a When the complexity of a model increases the *comprehensibility* of the model decreases, while its *predictive power* generally increases. **b** Since both the properties contribute to the *usefulness* of the model, we conclude that models of intermediate complexity are the most useful. Here we have assumed that usefulness is a product of the predictive power and the comprehensibility

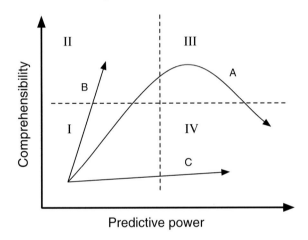

Fig. 5 Different classes of models ordered according to their predictive power and comprehensibility. If we classify models according to their predictive power (x-axis) and comprehensibility (y-axis) then each model corresponds to a point in the two-dimensional diagram. Naturally one strives for both high comprehensibility and predictive power (field III), but the preferences and disciplinary bias of the scientist also influence the position of the model in the diagram. A basic mechanistic model is easy to comprehend, but often lacks predictive power (field I), while a phenomenological or statistical model does not capture internal mechanisms, but solely aims for good predictions (field IV). In this diagram it is also possible to view the iterative process of model construction, here represented by arrows (A–C). If we start with a simple model in field I it is possible to imagine several different trajectories: (A) the comprehensibility and predictive power of the model both increase initially, but as the complexity increases the understanding of the mechanisms acting within the model decreases and hence the comprehensibility suffers; (B) mathematicians and physicists often aim for simple models and care less about predictive power; and finally (C) a model might also develop towards greater predictive power without providing any understanding of the phenomenon

viewed as "black boxes" that, given input data, generate predictions without capturing the true dynamics of the phenomenon (i.e. the structural similarity is low). In line with the above discussion we conclude that models in field III, which exhibit both high predictive power and comprehensibility, are superior.

Yet another application of this framework is to view iterated model construction within the same diagram. An early stage prototype could hypothetically end up in field I, and then move upwards (become easier to comprehend) and to the right (become more predictive) towards field III. However, there is a chance that the model becomes to complex and the comprehensibility goes down (trajectory A). In this diagram it is also possible to discriminate between different preferences and views on modelling. A mathematician with an interest in analytically tractable models would move along trajectory B, while someone working on a phenomenological or statistical model would strive for higher predictive power and move along trajectory C.

Generality and Diversity

There is a clear connection between the level of detail in a model and the generality of the conclusions that can be drawn from it. The appropriate level of generality might be difficult to achieve, e.g. if one tries to construct a model of a human disease, then it should not be specific to a certain human being, and at the same time not as general as to capture the dynamics of any human disease. In each case there is a balance between specificity and generality and where on the spectrum a model is placed often depends on the preferences of the scientist. To highlight this point we will briefly consider an example from theoretical biology, where mathematical models are used in order to study the process of speciation, the evolutionary process that leads to the formation of two species from a single ancestral species.[30] The dynamics of this process is a fundamental question in evolutionary biology. When constructing models of speciation minimal assumptions are made about the species in question, since the aim is to understand the general mechanisms driving this process. By establishing under which conditions (selective pressures, migration rates etc.) speciation occurs within the model, it is possible to compare the results with real closely related species and see if they match the criteria derived from the model. If there is a match then the model provides a possible explanation of the process, if not then the model lacks some crucial element and needs to be revised.

Models that are aimed at more general questions tend, for obvious reasons, to become more abstract. Possibly the most striking example of this comes from the Hungarian mathematician von Neumann. He was interested in one of the most fundamental questions in biology: What are the mechanisms and logic behind repro-

---

[30]Geritz, S.A.H., Kisdi, E., Meszéna, G. and Metz, J.A.J. (1998). Evolutionarily singular strategies and the adaptive growth and branching of the evolutionary tree. Evol. Ecol. 12:35–57.

duction?[31] Instead of studying reproduction in the biological world he reformulated the question into: Is it possible to construct a model in which reproduction occurs? By a model he meant a formal mathematical system, and to answer the question he formulated a completely new mathematical construct known as a cellular automaton. A cellular automaton consists of a square lattice, like a chess board, but with the difference that each square can be in a number of different states, usually represented by different colours (von Neumann used 17 states), and the state of each square changes depending on the state of the neighbouring squares. It is the states, together with the rules that govern the transition between different states, that define a cellular automata. von Neumann succeeded in finding a set of rules where a certain structure (configuration of squares in certain states) would over time give rise to an exact copy of itself, and consequently he had shown that reproduction was possible within a mathematical construct. The cellular automaton he devised was not a reflection of any existing real structure, but instead a representation of the basic phenomenon of self-replication. von Neumann's work was the starting point for a research discipline called "artificial life", which studies general properties of living systems, such as self-organisation, communication and swarming, from an abstract point of view. Models that are being studied in this field are not models of a particular system, but of abstract properties, but they are still to be considered models.

In contrast to theories, which are often considered as rivalling if they describe the same phenomenon, there is rarely such antagonism between models that overlap. This is because models often focus on different aspects of a system, and only if they assume a contradictory set of mechanisms do they disagree. This feature of models becomes clearer when we view models as tools: they might look different or be of different kinds (conceptual, symbolic etc.), but still achieve the same goal (describe the same phenomenon); or in other words, models are like different languages that portray reality in different ways. Because of this there is a pluralism amongst models that is not to be found amongst theories.

An example of pluralism among models is two different models that both describe the swarming behaviour of the slime mold *Dictyostelium discoideum*. This amoeba is a single-celled organism that resorts to collective behaviour when conditions are rough, such as during a drought, and this cooperative behaviour is in many ways similar to what is seen in multi-cellular organisms. When the organisms start to dehydrate they secrete a signalling substance that diffuses in the surrounding media. The substance triggers other cells to also produce it and at the same time move towards the direction of the signal. This gives rise to a spiral pattern of migrating cells, that in turn leads to an aggregation of cells at the centre of the spiral out of which a fruiting body is formed. This looks very much like a stalk and the cells located at the top form spores that can be carried away by the wind to new locations. The swarming of the cells was first described by the Keller–Segel model, a partial

---

[31] von Neumann, J. (1966). The Theory of Self-reproducing Automata, A. Burks, ed., Univ. of Illinois Press, Urbana, IL.

Fig. 6 Real and simulated
slime mold. The *upper panel*
shows aggregation of real
slime mold where the
fruiting body is starting to
form at the centre of the
colony (© Rupert Mutzel,
Freie Universität Berlin).
The *lower panel* shows the
result of an agent-based
simulation in roughly the
same stage of aggregation
(Palsson, E. and Othmer,
G.H. (2000), A model for
individual and collective cell
movement in Dictyostelium
discoideum, PNAS
97:10448–10453, © (2000)
National Academy of
Sciences, U.S.A.)

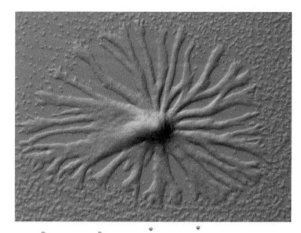

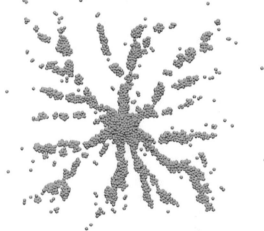

differential equation that describes how the concentration of slime mold cells (in the
unit cells/mm^2) changes in time and space.[32]

Since the model only describes the concentration of cells it can be viewed as
a relatively coarse-grained description of the phenomenon, and other models that
take into account the dynamics of single cells have also been formulated. In those
cases the system is described with a so called agent-based model, in which each cell
is modelled as a deformable visco-elastic sphere that interacts mechanically with
other cells and reacts to external stimuli.[33] Both models have their pros and cons: the

---

[32]Keller, E.F. and Segel, L.A. (1970). Initiation of slime mold aggregation viewed as an instability,
J. Theor. Biol. 26.

[33]Palsson, E. and Othmer, G.H. (2000), A model for individual and collective cell movement in
Dictyostelium discoideum, PNAS 97:10448–10453.

continuous Keller–Segel model can describe a large population of cells, and is also amenable to mathematical analysis, but does not allow for the tracking of individual cells. The agent-based model cannot handle as many cells, and is difficult to analyse, but could be utilised in order to investigate the effects of solitary mutant cells in a population of normal cells (Fig. 6).

There are also cases where a single model is able to describe two or more disparate phenomena. For example the Keller–Segel model has been applied in order to describe the formation of galaxies. To model this process one represents the stellar material as a self-gravitating Brownian gas, a large collection of particles that move according to a random walk and also attract each other according to Newton's law of gravity. A mathematical analysis of this situation shows that the density of stellar material also obeys the Keller–Segel equation.[34] This example shows that there are cases of two disparate systems that are governed by analogous mechanisms and exhibit a kind of universality. A connection can be seen to the mechanical analogies used by 19th-century physicists, where the microscopic world behaved *as if* it was made of springs and pulleys, but in reality it was not. A healthy degree of cautiousness is therefore recommended: just because two phenomena look alike doesn't imply that they are governed by the same mechanisms.

The Dangers of Modelling

By now it has become clear that all models are simplifications and idealisations of a much more complex world. This suggests that issues might arise when models are used and applied to real-world problems, and therefore in this section we discuss a couple of ways in which models might be misused and misinterpreted. This brief exposition will differ from the section "Models are always false, but is any thing at fault?" by focusing, not on mistakes made during model construction, but rather on problems that might arise when models are applied. These problems are more likely to occur when the person using a model is different from the person who developed and constructed it.

The conclusions that can be drawn from a model depend to a large extent on the simplifications that were carried out during its construction. As previously discussed, models are often created in order to capture and describe a particular situation and hence have a specific purpose. If taken beyond that purpose there are no guarantees that the model is useful and provides reasonable answers. The aim of a model is intimately linked to its construction and hence known by the person who constructed the model, but if the model is used by someone else then this tacit, but often important, knowledge might be ignored. This increases the chance that the model is misused and therefore provides invalid results. For example a model that is specifically designed

---

[34]Chavanis, P.-H. et al. (2004), On the analogy between self-gravitating Brownian particles and bacterial populations, Nonlocal elliptic and parabolic problems, Banach Center Publications, Vol. 66, pp. 103–126.

for making predictions on short time-scales might be utilised in order to make long-term predictions, or a model could be valid only in a certain parameter range, outside of which the approximations within the model are not valid. A concrete example of this are models of laminar flow of a fluid, which are valid only when the velocity of the fluid is small (or rather when the system is characterised by a small Reynolds number). If the velocity increases the characteristics of the flow change from laminar to turbulent, and this type of flow requires a different kind of model to be properly described.

Another more serious type of misuse occurs when a model is no longer perceived as a simplification, but is assumed to be a direct representation of reality. This situation is considerably worse since it not only might lead to false predictions, but also could lead to a completely wrong-headed conception of how the world is constituted. An instance of this is the model of the atom devised by Niels Bohr in 1913. This model can predict several characteristics of the hydrogen atom to a high degree of accuracy, such as its spectral lines, but it contains the invalid assumption that electrons travel in circular orbits around the nucleus, much like the planets of the solar system orbit the sun. The correct description in terms of the probability distribution of the electrons around the nucleus (atomic orbitals) was discovered in the 1920s by Pauli and Schrödinger, but the image of atoms as a miniature solar systems persists.[35]

In contrast to the above problem we have the reverse worry where models are looked down upon by scientists and laymen alike, simply because they contain simplifications. That an explanation of a phenomenon is based on a model is seen as a drawback, since the trust in simplifications of reality might be low. This issue has been brought to the fore in the debate about climate change, where it has often been claimed that the models utilised by climatologists are inadequate simplifications of a system we in reality know very little about. With regard to this it might be worth pointing out that although a model might not be able to describe and explain every aspect of a phenomenon, it is still better than not knowing anything at all.

Issues that arise in connection with models are enhanced if they are used in political or societal policy making. Conclusions drawn from models might e.g. affect how environmental problems are dealt with. This issue has been discussed by Orrin Pilkey and Linda Pilkey-Jarvis in the book *Useless Arithmetic* (2007), which focuses on environmental problems. Even though the book is critical of mathematical modelling, and unfortunately contains a number of misconceptions about the subject, it highlights a number of important topics. Certain phenomena, such a shoreline erosion, are so complex that we currently are unable to model them with the accuracy that society requires. In this case, should we use poor mathematical models that

---

[35] A contributing cause might be that Bohr's model of the atom is easier to comprehend and relate to compared to the quantum mechanical description, which doesn't describe any actual orbits, but instead provides a probabilistic description, which to each point in space assign a probability of finding an electron in that location.

provide wrong answers, or instead make use of conceptual and qualitative models, that do not provide exact answers, but at least are open to interpretation? Or in other words: what is the best model to use if we want to describe an unknown phenomenon: a mechanistic, quantitative model or a qualitative, phenomenological model based on experience? The answer is rarely clear cut, and depends on a number of factors, such as purpose, the existence of underlying theory, and the possibility of comparing the model with data.

Interviews

In the previous chapter we discussed and analysed the structure of models, their relation to theories and phenomena, and the basics of model construction. In this chapter we will be less concerned with philosophy and instead focus on the practical application of models. In order to achieve this we have looked beyond theoretical arguments, and gone straight to the source, namely to the scientists themselves. The reason being that it is their opinions and views on modelling that actually affect how and where models are constructed, and how they later are applied.

The interviews consisted of a number of questions related to models and their application, and were conducted over the phone. Our intention was never to quantify the answers in any way, and this brief survey should therefore be labelled qualitative. The sound recordings were transcribed, the interviewees were anonymised, and we only identify them by their respective discipline. The answers are presented, one question at a time, together with a concluding comment. We would like to highlight that the answers should not be viewed as representative for the discipline, but rather reflect the views of a single scientist.

What Is a Model?

Hydrogeologist: A model is a description with a certain purpose. Or in other words: a number of assumptions that are used in order to describe a system with a particular purpose in mind.

Mathematician: A model is an explicitly delineated structure that should describe a phenomenon in the real world in such a way that it does not blend with surrounding phenomena. One has isolated a chunk of reality and defined clear rules. Of course this is both positive and negative. The advantage lies in the fact that one can use the full force of mathematics in the model, but the down side is that the conclusions drawn are only valid within the model.

© Springer International Publishing Switzerland 2016
P. Gerlee and T. Lundh, *Scientific Models*, DOI 10.1007/978-3-319-27081-4_4

Climatologist: A model is a description of a system that one wants to study, that can be used in order to understand the system. With a model it is possible to gain knowledge that leads to further studies and in turn to more knowledge about the system. The meaning of the word is something that one has to keep in mind when talking to colleagues, since one might mean different things although we all use the same word. There are models that can be viewed as statistical relations between different types of data, or models that are highly complex and based on basic physical principles. In the latter case one typically starts with the laws of Newton together with thermodynamic laws and develops a model in the shape of equations together with a set of boundary conditions.[1] I work with physical models that are built from a basic description.

Ship-building engineer: A model is a representation of reality that makes it possible for us to understand it despite its complexity.

Astronomer: A model is a way to mimic some aspect of reality. The model is unable to fully describe the real phenomenon or object, but does so to such an extent that we can draw conclusions about the real object and preferably make predictions.

Zoologist: Within my discipline one often talks about animal models or model organisms, which can be used in order to draw certain conclusions and also extrapolate findings to other organisms. A wide array of organisms such as mice and fish are used. The biological research community has reached a consensus on which species to use, e.g. zebrafish is common when it comes to toxicological studies. Model organisms have certain properties that make them suitable, and in addition their genes have been mapped out. We also use computer models to study specific enzymes or cellular receptors. Such models can be applied in order to investigate how different chemical compounds interact.

Organic chemist: It depends. In my view there are three different levels of models. A model can be a mathematical relationship that correlates the chemical structure with the properties of a molecule. It can also be a guess of a structure provided one has three-dimensional information about similar structures, obtained from e.g. x-ray analysis. Lastly, a model can also denote the plastic constructions that we often use to visualise known chemical structures in order to identify bonding angles and van der Waal surfaces.

Economist: A model is a simplified representation of reality in a general formulation, often in mathematical terms, but not necessarily.

Neurologist: It is something that reflects the true state of nature, but that in addition can be manipulated.

Statistician: In statistics there are for example regression models that describe how the value of a variable is affected by predictors, i.e. the value of other variables called

---

[1]Authors' comment: Boundary conditions refer to the requirements and conditions that delimit a mathematical problem. In the case of bacterial growth in the introduction (p. 2) the carrying capacity L and the initial population $n(0)$ are boundary conditions.

independent variables. The model also includes a random element in terms of noise. This is however only one example of a statistical model, and generally speaking I would say that a model is a means to describe specific aspects of reality. When one creates a model there is often a trade-off between simplicity, with the goal of simplifying reality, and complexity, since the model needs to have some minimum level of complication in order to capture the relevant aspects of the system. It is the purpose of the model that dictates its complexity.

Comments

From the answers it is obvious that a model is viewed as something restricted. There is a certain purpose of the model and it describes/represents/reflects a limited part of reality. The goal of a model is (by being a simplified description) to create understanding of a phenomenon, but it is also used in order to make predictions. Yet another aspect is brought into light by the neurologist, the ability to manipulate models. The scientists roughly describe two different kinds of models: on the one hand mechanistic models that describe relations between different components within a system, in the case of physics with the aid of basic physical laws (e.g. Newton's laws), and on the other hand models that are based on statistical relationships and correlate variables within the system (phenomenological models).

How Do You Use Models in Your Research?

Hydrogeologist: I always set out to build a model, at least that is my goal. If it's mathematical or not depends. If it is mathematical I need boundary conditions, data, processes, representations etc. I'm often asked: "Can you describe this rock?", which is a perfectly valid question, but I also need to know why they're asking, what their purpose is. Which processes are they interested in? Is it flow of water, heat conduction, or deformation? These cases require different descriptions. If I say that the rock is red and two million years old, that is a true statement, but not necessarily the answer they are looking for.

Mathematician: I don't. It's a bit paradoxical that all scientists use mathematical models except pure mathematicians.

Climatologist: In my area of science the system under study is extremely complex. We are unable to measure every aspect of the system so instead models are used as a means to understand these unmeasurable relationships. Models are also used in order to make predictions of the future, for example to investigate different scenarios, and analyse how robust the predictions are, a method termed sensitivity analysis. Models help us towards understanding the system, and can also be used for supporting policy making in relation to a variety of environmental issues.

Ship-building engineer: In order to study the phenomena I'm interested in I have to use numerical models. Apart from these computer-based models we also use physical scale models. I'm not personally involved in the experiments carried out in the model basin, but I use the results as a comparison to results obtained from my simulations.

Astronomer: There's always a particular thing I'm interested in when I construct or make use of a model. We use models both as way of representing physical objects, such as galaxies, and as a way of illustrating what happens when certain properties of the objects are altered. So I use models either in order to depict an actual object, or as a means of illustrating changes to that object.

Zoologist: We use a species of fish, rainbow trout, as our model organism. For that particular species we have a lot of background information from previous experiments, and we have also developed tools, such as biomarkers, which of course are species specific. The reasons why we have decided to work on rainbow trout is that it is easy to acquire from fish farms, and it survives in our aquariums. Other researchers who are interested in the behaviour of fish might pick a different species, all depending on what type of behaviour one is interested in. The choice of a model is thus influenced both by practical reasons and by the scientific question under study. I also work with another type of model, a cell line, which really is a population of cancer cells. It is cultured in vitro in a petri dish and by using this model system, hypotheses can be tested and conclusions drawn before we move on to experiments on whole animals.

Organic chemist: I use three different kinds of models in my work: as a mathematical relation which connects chemical structures and properties; as a qualified guess of a structure given the three-dimensional information of similar structures; or as plastic models of molecules which help us visualise chemical structures and hence reveal their functionality. In teaching I mainly use models in the third sense, while in my research I use models in the second sense, i.e. as initiated guesses of molecule structure. In terms of proteins, which I work on, this initial guess together with the known amino acid sequence of the protein is fed into an algorithm, which is based on a mathematical model of protein folding. The algorithm then gives us a three-dimensional structure which is more accurate than the initial guess. Lastly, in pharmacological research, models that correlate molecular structure and properties are quite common.

Economist: I use models both in terms of simplified representations which allow for intuitive insights about mechanisms acting in the complex reality, but also as a means of generating and testing underlying hypotheses. Sometimes these tests can be quantified, although this is not necessarily so. Some type of test is required, however this can sometimes be achieved through introspection alone, but most commonly empirical tests are used.

Neurologist: I use them extensively in my work. In my line of research we make use of actual cases, patients with a particular disorder, not as models, but as the key. We study certain aspects of these patient cases, which are measurable in cell lines or experiments with zebra fish, and use these model systems as a means of investigating the dynamics of the disorder in the patient, which is what we really care about. An important aspect of our research is to establish that a model actually represents what we think it does. This is the key to translational research, that the model actually reflects the dynamics of the real thing.[2]

Statistician: I use for example regression models (see previous question). An actual example of such a model is one which I use for identifying genes that influence the risk of getting a stroke. In that model we consider the genes as having an impact on the risk, but we also need to take into account other important factors that influence the risk of stroke, such as age, smoking, physical activity, body mass index etc.

Comments

The answers to this question clearly reveal the differences between the disciplines, although some interesting points of contact are present. Many of the scientists use models as a substitute, since it isn't possible to acquire data from the actual system, or because the actual system is difficult to manipulate. We also see an interplay between different levels of models; the ship builder uses scale models as a way to validate numerical models, and the chemist lets an initial guess (a conceptual model) interact with a symbolic model in the shape of an algorithm, while the zoologist and neurologist use cell lines to model processes that occur within model organisms (rainbow trout and zebra fish), which in turn are models of humans or other species. An important distinction that emerges is the one between prediction, which the statistician, climatologist and chemist refer to, and the understanding of a particular system, brought up by the economist and astronomer.

What Makes One Model Better Than Another?

Hydrogeologist: It depends on a number of things, for example on the available data. We often split projects into stages. Early on we often have little knowledge about the system, and in that stage it's useful to have a model than can handle a smaller amount of information. Later on we extend the model and take into account other aspects. But there is no point in having a super-complicated model from the very start, since we cannot supply it with the data it requires. If presented with two different models at the same stage, then I would say that the model that answers the question posed most accurately is the best model.

---

[2]Translational research is an interdisciplinary method which aims at transferring findings in basic research to the clinic. Typically one starts with an experimental phase, then moves onto animal models and finally to patients. The results are then evaluated and the whole process can be repeated. Compare this circular process to the figure which describes model construction on p. 40.

Mathematician: It's difficult to say since the question is two-dimensional. On the one hand we have complex models that try to cover as much as possible. These are in some sense closer to the truth, but on the other hand they are difficult to handle and the answers they provide are not as clear cut. If one simplifies too much the answers are clear and neat, but the model might not describe the real system.

Climatologist: If we look at a specific question then the most important thing is how much a model deviates from other models, but this doesn't really tell us which model is better. Even if a model deviates from others it might still give an accurate description. One option is then to evaluate the model and compare it with independent data that was not used to create the model.

Ship-building engineer: The answer goes two ways, since it is desirable to have a large scope, but this often leads to reduced accuracy. I would say that a good model is one that is validated and has sufficient scope, and by scope I mean that the model is applicable in different contexts.

Astronomer: In terms of the models that I use I look at specific properties of a model, for example does it conserve energy on large time-scales, or does it conserve angular momentum? Does it reflect certain laws of physics in a better way or over longer time-scales? Is the resolution in time higher? Are the dynamics of the model more similar to reality? Can it describe a larger number of stars? Is it possible to include gas? When I started working with models of galaxies we could simulate the movement of almost 100,000 stars, while a real galaxy in fact contains roughly 200 billion stars. This number is still beyond our capabilities, but today we can simulate a couple of million stars. In a way this is an improvement, although still quite coarse-grained compared with a real galaxy. I would say that a model can be superior in three different ways: it agrees better with experiments and observations, it is more consistent with the laws of physics, and it makes it possible to study something that was previously inaccessible.

Zoologist: A lot of things. If we look at a cell line that consists of transformed cells originating from a tumour grown in a lab, then these cells have probably lost certain key properties. This is something we might suspect when the cells don't behave as expected. This is an important aspect when deciding on a model. Outside of the lab, when working in the field, there are other things to keep in mind. If one studies fish, then it is preferable to look at a species that is stationary, that doesn't travel very far. If one studies a species of fish that swims about over large distances then it's difficult to know where they've been and what they've been exposed to. The habits of the species also affects the suitability, if they are herbivorous or carnivorous, if they swim close to the bottom or are pelagic like a cod. These are a couple of things that we need to consider when we pick a model organism.

Organic chemist: If the model can provide accurate predictions.

Economist: If the model generates better, or more important, insight about the under-lying complex reality. This doesn't mean that it is more correct in some absolute sense. A perfect representation of reality is completely accurate, but immensely complex and doesn't explain a thing. There is always a trade-off between complexity and simplification. In economics there are different views on what constitutes a good model. For example some claim that a model is good if it produces accurate predictions even though the mechanisms it contains are flawed. In this case it's simply the ability to make predictions that decides the quality of a model. I'm not of that opinion myself. It's easy to paint yourself into a corner with that type of reasoning. I claim that there's a point in including underlying realistic and correct mechanisms in a model although the predictions are not as good.

Neurologist: If the model reflects the true state of the system to a high degree, and in addition is manipulable. Since the long-term goal is to develop new drugs it is of importance that one can manipulate the model.

Statistician: It is difficult to tell since it's a mix of mathematical aspects, where a model that can explain more of the variation is considered better, and pragmatic standpoints that we encounter when trying to predict the risk of contracting a disease. For example we would like the predictors to be accessible and therefore prefer a blood sample over a sample from the spinal cord. Even though the blood sample is less accurate, it's simpler, cheaper and safer to obtain compared with a sample of the spinal cord. Another pragmatic aspect is the ability to understand the model. A model that is too complex from a mathematical point of view, that is only understood by the statistician, is unsuitable since the people that will make use of the model don't grasp the underlying mathematics.

Comments

The question of the value of a model is obviously difficult to answer and many of the scientists consider it to be a trade-off between the comprehensibility of a model and its ability to make accurate predictions. Also practical details matter, as is evident from the answer given by the zoologist, and also the applicability, as is highlighted by the statistician. The chemist and neurologist are mainly interested in accurate predictions, and the latter also points to the importance of manipulability. For the hydrogeologist the dependence on context is large, and the value of a model depends on the stage of model construction. Yet another aspect is brought to light by the ship-building engineer, namely the fact that models with a larger scope, i.e. models that are more general, are preferred, but that this is often in conflict with the accuracy.

Have You Ever Met People with Ideas Different from Yours When It Comes to Models and Modelling?

Hydrogeologist: Yes, often. My answers so far have related to conceptual models that I view as a collection of assumptions about a system. What processes are important, what kind of data are necessary to describe these processes? In addition to that we need a law or equation, such as Darcy's law,[3] that defines the problem. Some scientists view conceptual models merely as superficial descriptions, while I find them detailed enough to transform them into computational models. In my opinion a conceptual model should contain enough detail so that one easily can make the model computational. It shouldn't just be a vague description. I have often encountered these different views on conceptual models among colleagues at different universities.

Mathematician: A former student of mine, who decided to move into more applied research, once told me: When a scientist approaches a mathematician with a problem she often gets the reply "I wasn't able to solve your problem, but I have found a very neat solution to a closely related problem".

Climatologist: I once talked to a colleague who works in a different field and told him that I work on models that are based on physical principles, and in return got the harsh reply: "Models are not physical!" The person worked on statistical models, or models that are based on statistical relationships. At that point we didn't have the possibility to reach an understanding since our premises were so different. In my view there is a fundamental difference between mechanical and statistical models.

Ship-building engineer: No, I don't believe so. I have rarely discussed the concept in itself. It's taken for granted that we all work with models. The people I've met usually work with models similar to mine, and use them in similar ways. In fact I've talked to a meteorologist who had the exact same view on models and the underlying physics as me—and who even used the same equations.

Astronomer: Sometimes there is a clash of cultures when someone says "but that's only a model". There's something negative about that statement–that the model can be discarded simply because it isn't a realistic representation of reality. I've encountered that attitude, and when that happens I try to explain in more detail what a model actually is, and what it can be used for. History has shown us that when a model is proven to be incorrect it doesn't follow that it's useless, like when the Newtonian worldview was replaced by Einstein's theory of relativity. The view that "if we don't know exactly how something works then there's no point knowing anything" is negative, and by no means constructive. When it comes to mathematical models it varies a lot across disciplines how much they are applied, although I believe they could be used more extensively.

Zoologist: Some people dismiss the use of single cells grown in a lab as means of drawing conclusions about the dynamics of an entire organism, since the cells have been taken out of their physiological context. The kidneys might affect the blood pressure that in turn affects the heart. If one organ is perturbed on a certain level then it might influence other organs, nerves and hormonal levels. If the physiological context is crucial then the results obtained from single cells are usually not trusted— the results might be in vitro artefacts. Similar arguments can be applied to genetic

---

[3]Authors' comment: Darcy's law describes flow in a porous material such as the slow trickling of ground water through soil.

experiments in single cells. It isn't reasonable to simply remove a single gene, study the effect of this in a cell line, and from that draw conclusions about its role in an organism. The physiology is very important, and other mechanisms, not captured in the cell line, might compensate for the loss of the gene. Because of this it's important to interpret your data carefully, and to clearly state in which system the conclusions apply. In my opinion there is room for both studies on cell lines and model organisms, such as mouse models, but one has to be careful about communicating the results. Meanwhile there are people that find studies on cell lines pointless.

Organic chemist: No, I can't recall that I have. We've often had differing opinions about details, but not on our general view of models. The reason for this might be that I've only collaborated with other natural scientists and we all shared the same views on modelling, or there maybe was some misunderstanding that never surfaced.

Economist: Yes, I have. Some people think that models only belong in the natural sciences, where matters in some sense are less complex and easier to delimit. Some social scientists find it provoking to construct a mathematical model of human behaviour. That point of view is not uncommon.

Neurologist: Yes, it's quite common to have an exaggerated trust in your own model. This might be a model organism, a cell line or even a chemical reaction that is claimed to exhibit some of the characteristics of the disease. Some scientists in translational medicine claim that they have a model of a certain disease, but in fact the model doesn't reflect the nature of the human disease, and therefore in my view doesn't qualify as a model.

Statistician: I used to work in software engineering. In that trade the discrepancies between model and reality are not as easily quantified as in statistics. Models in software engineering cannot capture all aspects of reality, and sometimes even tend to change it. The fact that models alter our view of reality is captured in the phrase "we can't do that because of the data", and is a consequence of the fact that computational models are in the driver's seat. Often one starts with a model in order to construct a computational system, such as a business system, and because of this the tables have been turned between model and reality. The model comes first, and after that reality. If I was to say anything about working with scientists from other fields then it would be that there is a lot of spread when it comes to previous knowledge of modelling. There is also variation in the level on which scientists want to describe and model their systems. In my opinion it's important to start with a top-down approach that gives a rough picture of the system. I have noticed that fast and successful collaborations often work this way. There are also scientists that immediately focus on minute details on the periphery, which makes it difficult to construct a good model.

Comments

The answers to this question highlight a number of issues that the scientists have encountered when it comes to modelling, both within their own discipline, but most commonly in contact with other disciplines. Firstly the hydrogeologist describes the problem that conceptual models are rarely viewed as proper models, rather as vague descriptions or preliminary attempts. The view of the mathematician is rather based on the differing ambitions that scientists might have when it comes to models. A mathematician often strives for a complete understanding of a model, while an applied scientist is more interested in the real system. A similar matter is brought up by the climatologist, who focuses on the difference between mechanistic and statistical models. Although this is more a matter of modelling approach, in essence the two situations are similar; mechanistic models aim at a more or less complete description while statistical models focus on the outcome and prediction. The astronomer describes a common view of models as "simply models" and also highlights the importance of false models. The economist has encountered a similar kind of scepticism against models within the social sciences, based on the view that certain phenomena, such as human behaviour, are too complex to be captured by simple mathematical models. The zoologist talks about differing views, within the discipline, on using cell lines, and the neurologist expresses a fear that the connection between reality and model is too weak, and that some scientists have an exaggerated belief in their own models. Lastly the statistician highlights the fact that in some disciplines models are superior to reality, which seems problematic.

What Is the Difference Between Theory and a Model?

Hydrogeologist: The concepts overlap, a bit like the chicken or the egg. The model forms the basis for theory, but theory is possibly larger.

Mathematician: I'm a bit uncertain about this. I know what theory means in mathematics, but for example look at the theory of relativity. In my view there is very little difference between theory and model. In the social sciences theories tend to be more general and broadbrushed.

Climatologist: In my opinion, theories are generalisations and syntheses of acquired knowledge. Models are a way of testing theories, but also a means of extending theory. Models can also be thought of as data or data acquisition, either in a laboratory or in reality. Experimental measurements and modelling are two scientific methods that complement each other. Theories are generalisations that gain inspiration both from measurements and modelling.

Ship-building engineer: I tend to view models as more focused and applied. This might also be true for theories, so I don't really have a good answer to this question.

Astronomer: My immediate answer is that I use models in order to construct theories. In the process of constructing a theory I make use of a number of different models, some of them in order to formulate the theory, others to test it.

Zoologist: To me a model feels more practical than a theory. A theory is something that one tries to prove, and in doing so one might use a model. I don't view models as theories, but rather models as tools. For example, say that I have theory of how the liver metabolises some substance. How do I show that it's true? Well, I build a model.

Organic chemist: A theory is based on knowledge, while a model rests on hypotheses. I would say that a theory has more support than a model.

Economist: Theories are more comprehensive than models, at least that's my immediate thought. Theory is a wider concept. A theory might encompass many models, but a model cannot contain many theories. A theory is not necessarily a simplified representation of reality, something a model must be.

Neurologist: Models can be used in order to challenge theory. If the challenges are serious enough then the theory needs to be revised. In some way models test theories.

Statistician: In statistical theory there are numerous propositions and theories, for example the central limit theorem. Statistical theory is built by formulating and proving theorems, but this has nothing to do with models. Theory is built on abstractions, while models are only useful if they reflect reality.

Comments

From the answers provided to this question it is possible to distinguish a couple of different views on the relation between theories and models. The hydrogeologist and mathematician say that they are difficult to separate, and in some cases they might even overlap completely. Another view, held by the economist, is that theories are more general and that they can contain several models. A dynamic view is provided by the astronomer and climatologist, where models are used in order to construct hypotheses and build theories, but also to test theories, a fact that is also underscored by the neurologist and zoologist. Moreover the chemist describes models as hypotheses and theories as solid knowledge. Models can turn into theories if they are properly supported and verified. For the statistician theories represent the abstract basis onto which realistic models can be built, and from this perspective they are clearly separated.

Concluding Remarks

It is important to bear in mind that these interviews were carried out with a select group of scientist from a small number of disciplines. In addition to this we have only interviewed one scientist from each field, and hence the conclusions that can be drawn from this study are limited. Instead we have to view the results as a random sample taken from a large research community, but despite this a few important things can be said.

It is obvious that models play a central role in the work of these scientists. They have clear views on modelling and manifest ideas about the role of models in their respective discipline. At the same time it is evident that the concept of a model differs between the disciplines: partly in the view of what can be achieved with a model, and partly in the range of models that are applied. Hence we can conclude that there is a wide array of models are that being used in contemporary research.

The fact that models play a central role in all disciplines, and at the same time the actual meaning of the concept differs greatly, suggests the immense scope of the concept. Despite the fact that all the scientists mean different things by the word "model" there is something essential that binds all its uses together. The models provide access to a reality that in almost all cases is of such complexity that it cannot be described or controlled with-out some prior simplification and abstraction. And it is in the sense of a tool, which allows the scientists to approach and investigate reality, that the different meanings coincide and the similarities become most obvious.

Worked Examples

In this chapter we give a more detailed description of a couple of scientific models, and in addition an insight into how these models were constructed and the impact they have had in the field of research they belong to. The aim is to cover all the types of models that were presented in the model taxonomy (p. 28) and also to look at models that belong to the fields represented by the scientists in the previous chapter.

The first example is taken from chemistry and describes a symbolic/mathematical model that has a mechanistic derivation. The subsequent example, taken from hydrology, is also symbolic, but in this case it is a completely phenomenological model without any structural similarity. Yet another model, aimed at prediction rather than understanding, is presented in the third example and comprises analysis and prediction of time series. We then move on to an example of analogous modelling, where springs and dashpots serve as models for the biomechanics of single cells. With the help of the Lorenz model we then show how a property of complicated weather systems can be distilled and studied in a simple symbolic model. The next example is drawn from astronomy and shows how a conceptual model, in all its simplicity, can aid scientists to form hypotheses and theories about the life and death of stars. The two concluding examples are iconic models taken from neurology and ship building, with which scientists and engineers can study systems where symbolic or conceptual models are inadequate.

Langmuir Adsorption

Many chemical reactions are facilitated by the presence of so called catalysts, substances that take part in the reaction but are not consumed themselves. The mode of action of a catalyst varies, but the end result is that the energy of activation of the reaction is lowered, which leads to an increased reaction rate. The catalysis of chemical reaction is essential for living beings. Most people associate catalysts with cars, where they aid in cleaning the exhausts, but as a matter of fact nearly all of the

P. Gerlee and T. Lundh, *Scientific Models*, DOI 10.1007/978-3-319-27081-4_5

chemical reactions that take place in our bodies are controlled by a certain type of proteins, called enzymes, that serve as catalysts of biochemical reactions.

Catalysis is also an important tool in the production of chemical products. For example, iron is used as a catalyst when the elements nitrogen and hydrogen react to form ammonium, and nickel is used in the production of margarine, in a process where hydrogen gas reacts with unsaturated fatty acids. These two catalytic reactions share a common feature in that they occur on the surface of the catalysts onto which the reactants bind and react with one another. The products of the reaction then disassociate from the surface and can be harvested.

In order to control and optimise reactions such as these it is important to understand how a chemical compound is adsorbed (binds) and desorbed (is released) from a surface, and how these processes are influenced by external factors such as pressure and temperature.[1] A complete description of these relationships would be immensely complex and formulated in terms of quantum mechanical wave functions, but luckily there are models that provide simplified descriptions that for most purposes are sufficient. A common model of this phenomenon is so called Langmuir adsorption, first put forward by Irving Langmuir at the beginning of the 20th century.[2] In its most basic form, which we will discuss here, it describes, with the aid of a mathematical formula, how much a surface can adsorb of a certain compound as a function of the concentration or partial pressure of the compound.[3] It is a symbolic model that relates a number of physical quantities to each other.

Let us consider a gas (A) that is in contact with a surface (M) and assume that the gas is in a state of dynamic equilibrium (see Fig. 1). By this we mean that despite the fact that gas molecules are constantly being adsorbed and desorbed at the surface, the mean number of molecules bound to the surface is constant. This type of relationship is usually written:

$$A(\text{gas}) + M(\text{surface}) \underset{k_d}{\overset{k_a}{\rightleftharpoons}} AM(\text{surface}) \tag{1}$$

where the reaction to the right occurs with rate k_a and the one to the left with rate k_d. These constants are known as the adsorption and desorption rates and describe how fast each reaction occurs.

In addition to the assumption about dynamic equilibrium we will also make the following simplifying assumptions:

1. The gas molecules are assumed to only cover the surface in a single layer, i.e. they form a so called monolayer.

---

[1] Please note the difference between adsorption and absorption: the former case refers to the binding on a surface, while the latter corresponds to the uptake of a substance in another gas, liquid or solid substance.

[2] Langmuir, I. (1916). The constitution and fundamental properties of solids and liquids. Part I. Solids. Journal of the American Chemical Society 38(11): 2221–2295.

[3] The partial pressure is the pressure a gas would exert on a surface if all other gases (such as air) were absent.

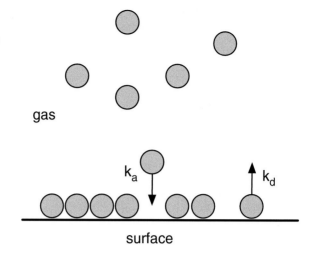

Fig. 1 A schematic image of a gas being adsorbed and desorbed from a surface. The constants k_a and k_d describe the rate with which the two processes occur

2. The surface is assumed to be homogenous and uniform. Hence we assume that it is irrelevant where on the surface a gas molecule is bound.
3. We assume that there are no interactions between the bound molecules. The likelihood of a molecule being bound in a specific location is therefore independent of the state of the neighbouring locations.

Let us now consider a part of the surface that contains N (independent) locations where adsorption can occur, and let θ denote the fraction of locations that are occupied. The gas A is adsorbed onto the surface with a rate that is proportional to the partial pressure of the gas p (this is because the number of collisions occurring per unit time is proportional to the pressure). But the rate of adsorption is also proportional to the number of unoccupied locations given by $(1 - \theta)N$, and lastly to the adsorption constant k_a. Therefore the total rate of adsorption is given by $k_a(1 - \theta)Np$. The rate of desorption on the other hand is given by the number of bound molecules θN multiplied by the desorption constant k_d. At dynamic equilibrium these quantities must be equal, which leads to the relation

$$k_a(1 - \theta)Np = k_d N\theta \tag{2}$$

which, with a bit of manipulation can be rewritten as

$$Kp(1 - \theta) = \theta \tag{3}$$

where $K = k_a/k_d$. This expression can finally be written as

$$\theta = \frac{Kp}{1 + Kp}. \tag{4}$$

Fig. 2 The Langmuir
isotherm describes how large
a fraction of a surface is
covered by adsorbed
molecules as a function of
the pressure p in the gas.
$K = k_a/k_d$ is the ratio of the
adsorption and desorption
constants

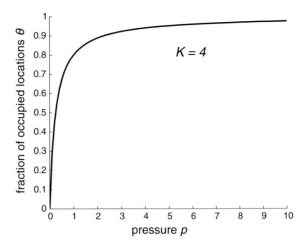

This relationship is usually called the Langmuir isotherm, and describes the fraction
of the surface that is covered θ (when the temperature is constant) as a function of
K and the pressure p. Figure 2 shows it as a function of p when $K = 4$, and we
can see that for small values of p the number of occupied locations increases almost
linearly ($\theta \approx Kp$), while when the pressure is large ($Kp \gg 1$) θ is roughly constant
and equal to unity.

This very simple model with its strict assumptions works surprisingly well and
is often used as a first approximation. It can also be extended in order to describe
catalytic systems where two or more gases are adsorbed to and react on a surface,
and is then referred to as the Langmuir–Hinshelwood model. There are also more
advanced models, such as the BET-model that allows for multiple layers of adsorbed
molecules and therefore provides more accurate predictions at high pressures.[4]

Rainfall-Runoff Models

There are probably few other things that are as important to us humans as water.
Every person drinks roughly two litres of water per day, but we are also dependent
on water in a number of technological contexts: as a solvent, a coolant and as a means
of transport, not to mention the food that we expect lakes, rivers and oceans to deliver.
The lack of clean water is a major problem in many countries today, a problem that
is likely to increase in the coming years. It is therefore important to understand the
natural processes on planet Earth that involve water, and as you probably know these
are highly dynamic processes. Water rarely stays in one place, but rather is part of

[4]Brunauer, S., Emmett, P.H. and Teller, E. (1938). J. Am. Chem. Soc., 60, 309.

constant cycle of rainfall, flow and evaporation. These are often called hydrological processes, and the discipline that studies them is called hydrology.

Many of these processes are however highly complex, composed of subprocesses on varying spatial and temporal scales, and in addition they exhibit large variation across space and time. Furthermore the underlying laws of physics that govern the processes are non-linear, which makes the systems sensitive to errors in initial data and parameter values. Taken together this means that fundamental physical models of hydrological phenomena are difficult and sometimes impossible to construct and analyse.

An example of such a phenomenon that is difficult to model is the relationship between precipitation and the flow of water in the watershed that drains an area. This relation is not only of theoretical interest but also of major practical concern, since for example for farmers it is important to know how much rainfall will reach their fields and when it will do so. At first sight this problem might seem trivial: large precipitation results in a high flow, but on a closer look it turns out that the problem is much more complex. The flow not only depends on the recent amount of precipitation, but also on the dampness of the soil, the geomorphology of the area, evaporation, infiltration and the intensity of the rainfall.

But if physical models are insufficient for predicting water flow, what other alternatives are there? One option is to disregard the underlying laws of physics, and hence any kind of mechanistic understanding of the process, and instead focus completely on prediction. That is to say we move from a model with a large degree of structural similarity to a completely phenomenological model. When taken to its extreme this type of model can be viewed as a black box, which is fed with initial data and produces a prediction, without providing any sort of understanding. One such class of models, which have had a large impact in hydrology, are so called artificial neural networks (ANN).[5] This type of model was originally inspired by the construction of the human brain, which has an amazing ability to recognise patterns and make predictions, and has spread from computer science and machine learning into many other disciplines.[6]

There is a wide array of different types of artificial neural networks, and we will focus on the multi-layer perceptron (MLP). It consists of a number of nodes or neurones that just like the neurones in our brains, are connected to each other in a network. Each node receives signals from a number of other nodes, and also has the ability to emit signals. These can be of two different kinds: either excitatory or inhibitory, but also of varying magnitude. Similarly to real neurones it is assumed that a node only emits a signal, or "fires", if it has been stimulated above a certain threshold value. If a node receives signals from more than one node then these are added internally, using the convention that inhibitory signals are negative and stimulating signals are positive. We can formulate this mathematically using the following formula for the internal state of node i in the network:

---

[5]Haykin, S.S. (1999). Neural networks: a comprehensive foundation. Prentice Hall.

[6]This is an archetypal example of a model inspired by biology, so called biomimetics.

$$x_i = f\left(\sum_j W_{ij}x_j - \theta_i\right) \tag{5}$$

where $f(x) = \tanh(x)$ is a so called transfer function, W is a matrix or table that describes the connections between the nodes and θ_i denotes the internal threshold value of the node. Without going into too much detail we can say the value of the node x_i is given as a weighted sum of the values of the nodes it is connected to, and that this sum is fed through the transfer function $f(x)$. This function is displayed in Fig. 3 and shows that the node behaves just as we described; only if the total stimulation is above the internal threshold value θ_i is the node active ($x_i > 0$), otherwise it is inactive ($x_i \le 0$).

The matrix W contains information about the structure of the networks and how the nodes are connected. In the MLP-network nodes are organised into distinct layers: one in-layer, one or more hidden layers, and an out-layer (see Fig. 4). The nodes in the in-layer correspond to the variables, in the system we want to model, that are known, while the nodes in the out-layer correspond to the variables we would like to predict. The network functions in the following way: First, data is presented to the in-layer, i.e. the node values are set to the values of the known variables. With the help of Eq. (5) the values of the nodes in the hidden layers(s) are then calculated, and lastly the state of the nodes in the out-layer are computed. The values of these nodes comprise the prediction made by the network, and because of its structure this type of network is known as a "feed-forward" neural network, where information is fed from one side of the network to the other.

But how is the matrix W, that holds information about the connections, determined? Its values are determined by training the network on known pairs of in- and out-data that exhibit the pattern that the network is later to recognise. Initially the values of W are chosen randomly, and one by one the training pairs are presented to the network. In each case the network produces a prediction P', and since we know

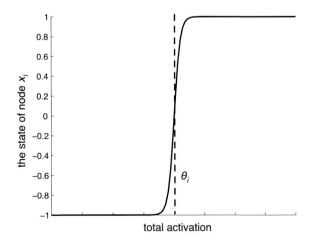

Fig. 3 The activity of a neurone in a neural network as a function of the input provided by the other neurones in the network

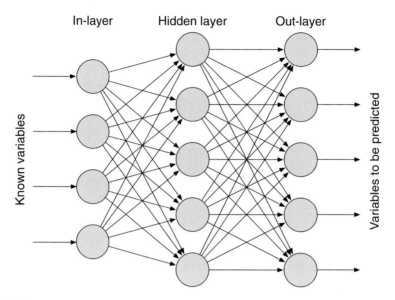

Fig. 4 The structure of a multi-layer perceptron (MLP)

the true out-data P we can calculate the prediction error $E = |P - P'|$. Using information about this error the values of W are updated in such a way that the network produces better and better predictions.[7]

Let us now return to the relation between precipitation and water flow. This system has been modelled successfully with the aid of MLP-networks. The in-data to the network is typically the precipitation at a number of locations on one or more occasions, and the water flow measured during a time-period prior to the date of desired prediction, while the out-data consists of the predicted flow in one or more locations. Figure 5 shows an example of this methodology, in which scientists have constructed an MLP-network that can predict the water flow in Eucha Watershed in Oklahoma, USA.

Unfortunately there is no mathematical theory that, given a certain prediction problem, prescribes the best size and structure of the neural network. Therefore it is up to the modeller to use trial and error (and experience) in order to find the optimal dimensionality of the in-data and the number of hidden layers, and the number of nodes in those layers. In addition, it turns out to be difficult to find the optimal amount of training data that balances learning with so called over-learning, in which case the network loses its ability to generalise to novel data. Despite these issues, neural networks are useful models when prediction is vital and an understanding of underlying mechanisms is viewed as less important.

---

[7] A common method for achieving this is "back propagation" that updates the values of the nodes in the different layers in sequence in an effort to minimise the error.

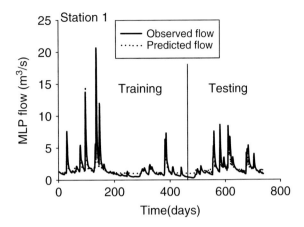

A Statistical Model for Time Series

How large will the future consumption of electricity be in Sweden? Of course this
question is practically impossible to answer accurately, at least in the longer term,
but if we could make a prediction, if only into the near future, it would be helpful.
Our inability to store electricity efficiently sometimes makes it difficult to match
supply and demand, resulting in rocketing electricity prices, something that could be
avoided if proper predictions could be made.

Let us for a moment go back to 1st March 2000, and based on existing knowledge
at the time try to predict the national consumption of electricity ten years ahead. At
our disposal we have information from Statistics in Sweden[8] in the form of a table
containing the monthly consumption measured in gigawatt hours (GWh), that starts
with January 1990: 15249 GWh, and continues with 13211, 14548, 12803, 11845,
10312, 9590, 10986, 12219, 14095, etc. A simple way to approach this type of data
is to plot it in a diagram with time on the x-axis and consumption of electricity on
the y-axis (see Fig. 6).

With the aid of this graphical exposition we can observe a pattern that infuses
us with some hope that a prediction of the future consumption is possible. The
most obvious feature of the consumption diagram is a yearly cycle.[9] In Sweden
more electricity is consumed in winter time due to the lower temperatures and lack
of daylight, while in countries close to the equator the situation is often reversed.
Another pattern that can be observed is a slow, but persistent increase in consumption.
This is most likely due to the "constant" economic growth, but we will limit ourselves
to the predictive power of the consumption data itself, and will not include any

---

[8]This is an authority that records and distributes a variety of statistics concerning Sweden. The data
analysed here can be found at: http://www.ssd.scb.se/.

[9]This feature can be analysed using an autocorrelation diagram, which quantifies positive and
negative correlations between different time points.

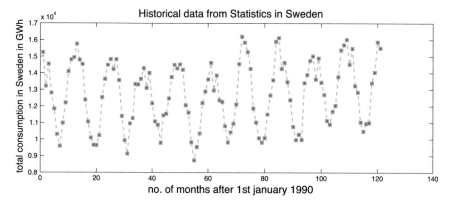

Fig. 6 How can we predict the future consumption of electricity in Sweden, based on the consumption over the last ten years? By plotting the monthly consumption in GWh and connecting the different time points a pattern appears: both a yearly cycle and a slow increase

external factors into the model.[10] A third observation is that there seems to be some element of "chance" to the consumption. This could for example be due to the seemingly random variation in temperature across different years, and during colder winters the consumption is likely to go up. Please note that we are not claiming that the weather and hence temperature is completely random, but rather that it can be modelled as if its influence on consumption is random.

How do we proceed and quantify the three different qualitative characteristics (cyclic, increasing, random component) of the time series? These questions are usually answered with the help of *time series analysis*. The yearly periodicity coupled to the slow increase makes it reasonable to assume the following mathematical model that contains a linear function for the increase and a trigonometric function with a period of 12 months for the seasonal variation:

$$y_m = \overbrace{a + bx}^{\text{linear part}} + \overbrace{c \cos\left(\frac{2\pi}{12}(x - d)\right)}^{\text{periodic part}},$$

where a, b, c and d are parameters to be optimised, or fit, with the help of the historic data. A common way to solve such an optimisation problem is to minimise the sum of the square of the error between the model and the historic data. If we let $y_h(x)$ denote the historic consumption at month x and $y_m(x)$ the prediction made by the model at the same time, then we are looking for a function y_m that minimises the sum over all time points

---

[10]If this restriction is relaxed then it would for example be possible to include the gross national product and consumer price index as input to the model.

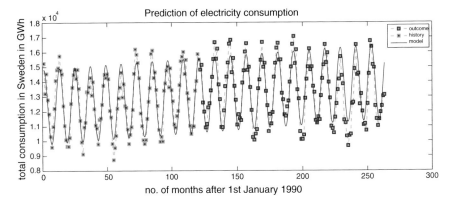

Fig. 7 This plot shows the three curves together: the historic data between 1990 and 2000 (*dashed line*), the outcome in the years 2000–2012 (*dashed line*), and the derived model (*solid line*)

$$E = \sum_{x=0}^{120}(y_h(x) - y_m(x))^2.$$

By applying this method we arrive at the following function as a candidate solution:

$$y_m = 12300 + 6.43x + 2517\cos\left(\frac{2\pi}{12}(x-1)\right).$$

With the famous quote by Niels Bohr, "It is very difficult to make predictions, especially about the future" in mind, we are curious to know how well our model matches the actual consumption data during the period 2000–2012. If we plot the consumption of electricity during the whole period 1990–2012 together with our candidate function we get the graph shown in Fig. 7.

Does the model provide a good description of reality? Visually it looks alright, but in order to get a better view of the agreement between model and reality we also plot the prediction error, i.e. the discrepancy between the real consumption and our prediction (see Fig. 8). This figure provides a clear view of the randomness that we discussed earlier, and seems to suggest that the deviations from a smooth consumption are indeed due to chance, which confirms our hypothesis that external factors can be modelled as if they are random. If there was some pattern in this figure then it would have been a clear indication that something was missing in the model.

It is worth pointing out that the model we chose assumes that the consumption follows an oscillating (cosine) function and that there is a linear increase, both of which are quite restrictive assumptions. More common methods in the field of time series analysis include ARMA- and GARCH-methods,[11] which smooth out the his-

[11]Mills, T.C. and Markellos, R.N. (2008). The Econometric Modelling of Financial Time Series, Cambridge University Press.

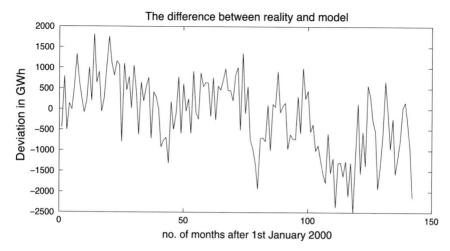

Fig. 8 How well did the prediction match the real data? This is answered by plotting the difference between the real outcome and the prediction made by the model. If the difference is centred around zero and random, then the model does a good job. In our case we find that the mean difference is −159, which is acceptable. However, the values are not normally distributed, which can be discovered by applying the so called Lilliefors test

toric data and contain a form of memory. Neural networks have also been applied in order to predict future electricity prices, and this modelling approach was described in detail on p. 70.

The Standard Linear Solid Model of Cell Biomechanics

Mechanical properties such as stiffness, stress and elasticity are often associated with the construction of roads, bridges and buildings, but are also necessary in order to understand how biological objects behave when they are stretched, compressed or twisted. With the help of biomechanics we can e.g. understand how to construct the best prosthesis and how a human body is affected by a car crash. But it might also be of interest to learn how things function mechanically on smaller spatial scales, such as the operation of single tissues (muscles, bones etc.) or even single cells. In fact modern technology makes it possible to investigate the mechanical properties of single DNA molecules.

In this section we will describe a model that captures the mechanical properties of a human leukocyte, a white blood cell, although the model could easily be generalised to other cell types. Leukocytes spend most of their time circulating in the blood flow, where they guard the body against unwanted guests such as bacteria. In order to understand how this transport, which is far from ordered and smooth, affects the cell it is necessary to have a basic understanding of how the cell behaves mechanically.

What happens to the shape of the cell when it is exposed to a force? Is it deformed, and if so how much? These are the kind of questions the model will be able to answer.

Despite its microscopic size, a human cell is an immensely complex object. It is built out of millions of molecules that work together and interact in a number of different ways. However, in order to achieve a basic understanding of the mechanics of a cell we don't have to take into account these molecular details, and we can focus our attention on the main components that affect the mechanics of the cell. The only things we have to account for is that the cell consists of a membrane and that this membrane is filled with a viscous fluid. The properties of the membrane, the fluid and the interactions between the two are in essence what determines the mechanical properties of the cell. In fact the membrane of a leukocyte is heavily folded, which implies that it affects the properties of the cell only to a small degree, at least as long as the deformations are small. The situation can be likened to a thin plastic bag that is half full and contains a viscous fluid, such as honey. If one pokes the bag it will deform, but the deformation will depend mostly on the properties of the fluid and little on the characteristics of the plastic bag. Therefore we will view the leukocyte as a homogeneous body that consists of a single viscoelastic material. A body that is viscoelastic is viscous like oil or honey and opposes deformation, but at the same time exhibits elastic properties like a rubber band.

In mechanical contexts one is often looking for a relationship between the stress inside a material σ (like pressure measured in N/m^2) and the resulting deformation or strain ε (a dimensionless number that measures the relative displacement). In terms of these quantities, if a viscoelastic material is exposed to a constant stress then the strain increases in time (known as creep), and if the strain is held constant, then the stress decreases with time (known as relaxation).

These properties are found in a simple model of viscoelastic materials known as the standard linear solid model, which only contains three mechanical components: two springs and a dashpot (see Fig. 9). The left-hand side of the construction is

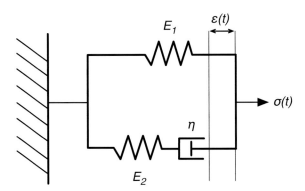

Fig. 9 A schematic image of the standard linear solid model of viscoelastic materials. The two springs have elastic moduli E_1 and E_2 respectively, and provide the construction with elastic properties, while the dashpot has viscosity η. When the construct is subject to a time-dependent stress $\sigma(t)$ it responds with a strain $\varepsilon(t)$ that also varies in time

anchored to a wall, and the right-hand side is subject to a strain directed to the right. The springs, which capture the elastic properties of the material, follow Hooke's law, and the stress is therefore proportional to the strain according to the relation:

$$\sigma_s = E\varepsilon_s \tag{6}$$

where E is the elastic modulus that measures the stiffness of the material. The dashpot, which captures the viscous properties, behaves according to:

$$\sigma_d = \eta\frac{d\varepsilon_d}{dt} \tag{7}$$

where η is the viscosity of the dashpot. This equation states that the stress in the dashpot is proportional to the rate of change of the strain, i.e. if the strain is constant then the stress is zero.

The connection between the standard linear solid model and the deformation of leukocytes is as follows: imagine that we subject the cell to a constant stress by using a micro-pipette to produce a negative pressure (see Fig. 10). The cell will react to this by deforming in a time-dependent manner. We want to relate the stress that the cell is subject to, to the strain that it exhibits. In the simplified model this corresponds to adding a stress $\sigma(t)$ and calculating the resulting strain $\varepsilon(t)$. With the help of basic mechanics and Eqs. (6) and (7) one can show that these quantities satisfy the following equation:

$$\frac{d\varepsilon(t)}{dt} = \frac{1}{E_1 + E_2}\left(\frac{E_2}{\eta}\sigma + \frac{d\sigma(t)}{dt} - \frac{E_1 E_2}{\eta}\varepsilon(t)\right) \tag{8}$$

where E_1 and E_2 are the elastic moduli of the springs and η is the viscosity of the dash-pot. If we assume that the cell is at rest for $t < 0$ and that a constant stress is applied $\sigma(t) = \sigma_0$ for $t > 0$ we get the following expression for the time-dependent strain:

$$\varepsilon(t) = \frac{1}{E_1(E_1 + E_2)}\left(E_1\left(1 + e^{-\frac{E_1 E_2}{(E_1+E_2)\eta}t}(\sigma_0 - 1)\right) + E_2\left(1 - e^{-\frac{E_1 E_2}{(E_1+E_2)\eta}t}\right)\right). \tag{9}$$

This relatively simple expression describes the behaviour of a real cell to a fair degree of accuracy (see Fig. 10). However, one needs to be aware of the limitations of the model. For example it does not work as well when the deformation is of such an extent that the membrane is being stretched or when the duration of the deformation is so long that the cell actively resists it.

By simplifying the original system, seeing similarities with another material and identifying the properties of that material with the properties of macroscopic mechanical systems (springs and dashpots), an analogous model was created. The cell is modelled as if it was made up of springs and dashpots, while in reality it has a completely different constitution. By making use of known properties of the components in the analogous system the behaviour of the original system can be described. In

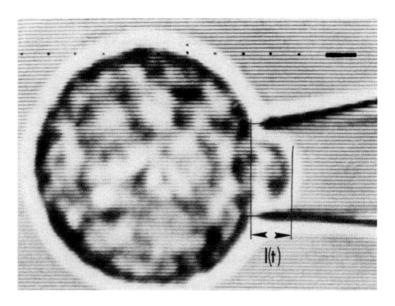

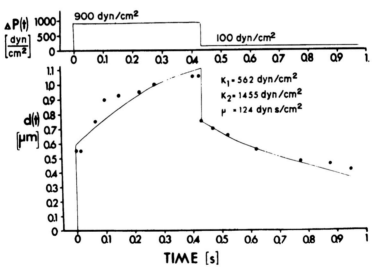

Fig. 10 The *upper panel* shows the experimental setup in which a leukocyte is subject to a stress produced by the negative pressure inside a micro-pipette. The *lower panel* shows how the strain varies over time as the pressure difference is altered. From a resting state the cell is exposed to 900 dyn/cm^2 which is then lowered to 100 dyn/cm^2 (1 dyn/cm^2 = 0.1 N/m^2). *Source* Schmid-Schönbein, G.W., Sung, K.L., Tözeren, H., Skalak, R. and Chien, S. (1981). Passive mechanical properties of human leukocytes. Biophys J. 36(1): 243–256, © (1981), with permission from Elsevier

fact the analogy can be taken one step further by comparing the spring and dashpot to electric components. If one equates the mechanical stress (pressure in the material) with electric pressure (i.e. voltage) and the time derivative of the strain with the flow of electrons (i.e. the electric current), then the spring corresponds to a capacitor that stores energy, while the dashpot corresponds to a resistor that transforms electrical energy to heat. With the aid of analogy we have gone from an extremely complex living cell to a homogeneous viscoelastic material, further on to a construction of springs and dashpots and lastly to an electric circuit.

From Weather to Chaos

The weather is, as we all know, sometimes difficult to predict. On shorter time scales, within the next couple of hours, it's in most cases possible to produce a reliable weather forecast, but when the prognosis stretches over a couple of days the uncertainty is much larger, and trying to make a prediction of what the weather will be like in ten days or more is almost impossible, at least with reasonable precision.

This inability to predict the weather on longer time scales is often taken for granted, but is far from obvious if you look closely at the underlying physics that controls the weather. The equations that describe how air and water are being transported in the atmosphere are fully deterministic, which means that they do not contain any element of chance. If the initial condition is known for a deterministic system, then, in principle, the future state of the system can be computed at any desired time point. But if the weather is deterministic how come it's so difficult to predict? The reasons are both practical and theoretical. A practical difficulty arises from the fact that we don't fully know the initial conditions of the system, i.e. the wind speed, direction of the wind, air pressure, humidity etc. at every point in the atmosphere, but only have the information about the conditions in those places where weather stations are located. Even though this problem might seem impossible to solve, the theoretical difficulties are if possible even worse.

The underlying equations that control the weather are non-linear, and this implies that solutions to the equations are extremely sensitive to disturbances in the initial conditions. This means initial conditions that are close to each other give rise to future states of the system that are very different from one another. It is often said that the initial conditions diverge from each other as the dynamics of the system unfolds. All physical measurements, including those made by weather stations, are made within a certain degree of accuracy, and this means that the uncertainties would grow over time, effectively making predictions arbitrarily far into the future an impossibility. This property of a system is known as *deterministic chaos*, and although the name was not coined until the 1970s, it is something that mathematicians and physicist have studied for a long time, e.g. in connection with the famous N-body problem.[12]

---

[12]The problem consists of solving Newton's equations of motion for N bodies simultaneously, where the force exerted on each body is the total gravitational pull from all other bodies. The

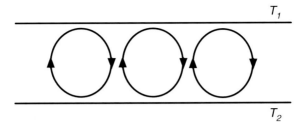

Fig. 11 Schematic image of Rayleigh–Bénard convection. When a thin layer of liquid is contained between two surfaces with a temperature difference of $\Delta T = T_1 - T_2$ convection cells might appear depending on the magnitude of ΔT

The concept was popularised by the meteorologist Edward Lorenz, who in the 1960s was the first to mathematically describe a hydrodynamic system that exhibits chaos. Based on his observations he also coined the phrase "the butterfly effect", which refers to the sensitivity to changes in initial conditions, and has its origin in the rhetorical question "Does the flap of a butterfly's wings in Brazil set off a tornado in Texas?"[13]

Lorenz studied so called Rayleigh–Bénard convection that takes place when a thin layer of liquid is heated from below and cooled from above (see Fig. 11). If the difference in temperature $\Delta T = T_1 - T_2$ between the top and bottom is small, then the fluid is stationary and the temperature of the fluid forms a smooth gradient from top to bottom. If ΔT is increased, so called convection cells appear, in which the fluid alternately rises and sinks in a circular fashion. This well-ordered structure is however destroyed if ΔT is increased further, and replaced by a turbulent motion without any sort of regularity, and this was the behaviour that interested Lorenz.

This system, at least superficially, resembles the dynamics in the atmosphere, where air is being heated close to the surface of the Earth, rises, is cooled down and falls to the ground. And if the idealised model system exhibits turbulent and chaotic behaviour, why not the entire atmosphere? Lorenz approached this question by simplifying the Rayleigh–Bérnard system with the prospect that the simpler system would also exhibit deterministic chaos.[14] Initially the convecting fluid was described by two coupled partial differential equations, one for the direction of flow

---

(Footnote 12 continued)
great French mathematician Henri Poincaré partially solved the three-body problem in 1887 and in doing so claimed the winning prize in a competition organised by the Swedish king Oscar II. In his solution Poincaré for the first time described the concept of deterministic chaos and laid the foundations for modern chaos theory. But the road to success was not straight. A mistake was found in Poincaré's original submission and he had to use all of the prize money to publish a revised version of his article in the journal *Acta Mathematica*, still one of the most prestigious journals in the mathematical sciences. For a detailed history of the N-body problem see for example F. Diacu (1996), The solution of the N-body problem. The Mathematical Intelligencer, 18(3).

[13]Lorenz, E. (1966). The Essence of Chaos. CRC Press.

[14]Lorenz, E. (1963). Deterministic non-periodic flow, Journal of the Atmospheric Sciences, 20, 130–141.

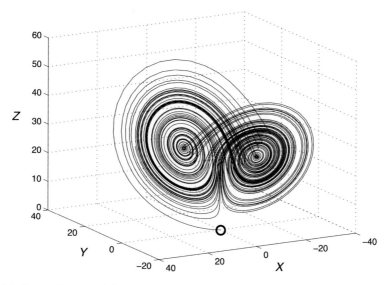

Fig. 12 A visualisation of the Lorenz attractor. The initial condition $(X(0), Y(0), Z(0)) = (0.1, 0, 0)$ is shown as a circle at the bottom of the figure

and one for the temperature. These equations are however not easy to handle, but by applying a clever transformation Lorenz was able to reduce the system to three coupled ordinary differential equations of the form:

$$\begin{cases} \frac{dX}{dt} = -\sigma X + \sigma Y \\ \frac{dY}{dt} = rX - XZ - Y \\ \frac{dZ}{dt} = XY - bZ. \end{cases} \tag{10}$$

Here X is proportional to the intensity of the convective flow, Y is proportional to the difference in temperature between the liquid flowing up and down, and Z is proportional to the deviation from a linear flow. The constants σ, r and b only depend on the physical properties of the liquid, such as the viscosity and thermal conductivity.

This system of equations might seem highly abstract and far from providing a description of the processes taking place in the atmosphere of the Earth, but we have to keep in mind that Lorenz tried to prove that weather systems can exhibit chaotic dynamics. If a simplified toy model does so, then it is very likely that the real system, with its many more degrees of freedom, also does so. And indeed he was right, at least when it comes to the toy model. The state of the Lorenz model (10) at a time t is described by the triple $(X(t), Y(t), Z(t))$, and the dynamics of the system can therefore be represented as an orbit through three-dimensional space, where the direction of motion at each point in space is determined by Eq. (10). Figure 12 shows such an orbit that was started with the initial condition $(X(0), Y(0), Z(0)) = (0.1, 0, 0)$

and reveals the complicated structure of the dynamics. This structure is referred to as the Lorenz attractor since it attracts orbits from all possible initial conditions. Because of its peculiar properties the attractor has gathered considerable attention from mathematicians. For example it has been proven that the Lorenz attractor is fractal.[15] The model also exhibits the desired sensitivity to initial conditions and has become a canonical example of chaotic systems.

We started out with weather and wind, and via convection ended up with the relatively simple Lorenz model that exhibits complicated and even fractal dynamics. The model does not have any predictive power, but instead makes it possible to explain a specific property of the weather, its sometimes unpredictable behaviour. Because of its properties the model is located at the outer edge of the predictive-explanatory spectrum of models: it cannot help us predict the weather, but instead gives insight into its basic dynamics.

The Hertzsprung–Russell Diagram

Claiming that the sun will rise tomorrow is a safe statement if there ever was one. Except for its seasonal variations the sun seems invariant, and it is only rare eclipses that darken its otherwise constant glow. It is therefore not surprising that many early cultures gave the sun a central place in creation, and often connected it to a mighty deity. In the Aristotelian philosophy, which dominated European thought all the way into the Middle Ages, the sun was placed in the celestial sphere that was viewed as immutable and close to perfection. This idea persisted into the 17th century when sun spots for the first time were systematically observed by Galileo Galilei and Thomas Harriot, which suggested that the sun could in fact change. Around the same time Tycho Brahe observed the first supernova, which added to the evidence in favour of the mutability of not only the sun but also the stars.[16] But if the stars are in a state of constant change, how are they born, how do they age and eventually die?

As scientific instruments for observing the sky improved, larger quantities and more accurate data were gathered, but it was not until the 20th century that astronomers could properly describe the life cycle of a star. A central tool in this discovery was the so called Hertzsprung–Russell (HR) diagram, which was developed during the 1910s by the Danish astronomer Ejnar Hertzsprung and Henry Norris Russell at Princeton University, independently of one another.

Around this time astronomers had realised that the colour of a star, or rather its spectral properties, are connected to its temperature. According to a classification developed at Harvard under the supervision of Antonia Maury, stars are organised into seven spectral classes, depending on the surface temperature of the star. The

---

[15]Tucker, W. (2002). A Rigorous ODE Solver and Smale's 14th Problem. Found. Comp. Math. 2: 53–117.

[16]The observation was made in 1572 at Herrevad Abbey in what is now Sweden, and the discovery was described in the writing *De novo stella* (1573).

temperatures vary from 3000 to 30,000 K and the classes are denoted (in decreasing order of temperature): O, B, A, F, G, K, M. Each class is subdivided into ten subclasses with the help of the numbers from 0 to 9. In this classification the sun is a G2-star with a surface temperature of 5800 K. In addition it had become clear that stars vary greatly in their brightness or magnitude. In one way this is obvious since they are located at different distances from the solar system and therefore appear more or less bright on the night sky. But what we mean is not the apparent magnitude, but the absolute magnitude of a star. This is defined as the apparent magnitude a star would have if it was placed at a distance of 10 parsecs from the Earth.[17] It might seem paradoxical that the absolute magnitude of a distant star can be determined, but to help them astronomers had a number of overlapping methods that allowed for accurate calculations of magnitude.

Spectral class and magnitude were essentially the only properties of stars that astronomers could observe and make use of when formulating theories of the composition and evolution of stars. An important step towards such theories was to investigate the relationship between these two properties. In other words, how do spectral class and magnitude relate to each other?

It is precisely this relation that the HR-diagram illustrates. In the two-dimensional diagram each star is placed according to its spectral class (on the x-axis) and absolute magnitude (on the y-axis) and what emerges is an obvious pattern (see Fig. 13). The dots in the diagram that correspond to stars are not randomly distributed, but most of them fall on a diagonal band, called the "main sequence". Above this band we find a collection of stars known as "red giants" and "super giants", and at the bottom left we find the "white dwarves".

This pattern was a significant discovery in itself, but the most important consequence of the HR-diagram was that astronomers received a powerful tool for exploring stellar evolution. If one follows a star through its history it can be traced as a curve or orbit within the HR-diagram. Hence astronomers now had a way to illustrate the problem and reason about it. An early example of this was Russell's own hypothesis of how stars live and die.[18] He thought that a star is formed through the gravitational contraction of a cloud of gas, and that this leads to an increased temperature which makes the gas glow. Further contraction makes the gas even hotter and this corresponds to a movement to the left in the HR-diagram. When the star reaches the main sequence it starts to cool, but keeps on contracting. Since its size is reduced, the magnitude is lowered, and coupled to the falling temperature this leads to a movement downwards along the main sequence (see the dotted line in Fig. 13).

This hypothesis was put forward in 1914, before it was known that it's nuclear reaction in the interior of stars that provides the stellar energy, and was instead based on the assumption that the emitted energy of a star is derived from the potential energy of the collapsing gas. According to the current view of stellar evolution the

---

[17] 1 parsec ($\approx$3.26 light years) is the distance at which a star exhibits a yearly angular displacement or parallax of one second of arc.

[18] For a detailed account of this and other examples we refer to: Tassoul, J.-L. and Tassoul, M. (2004). *A concise history of solar and stellar physics*. Princeton University Press.

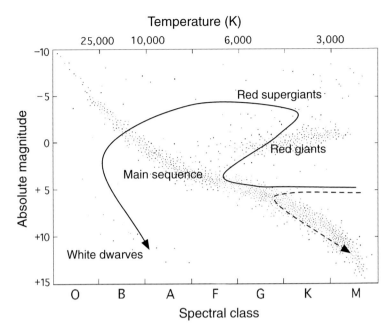

Fig. 13 The Hertzsprung–Russel diagram that shows the relation between the brightness or absolute magnitude and the temperature or spectral class. Most stars appear on the so called "main sequence", above it we find the giant stars and below the white dwarves. The *dotted line* corresponds to Russell's initial hypothesis about how stars move in the diagram when they age, while the *solid line* shows the current view of stellar evolution (see the text for details)

journey of a star in the HR-diagram looks slightly different, and can be summarised as follows: A cloud of gas slowly contracts and its temperature is increased; when a critical temperature is reached nuclear reactions are initiated in the interior of the star. This corresponds, just as in Russell's hypothesis, to a movement to the left into the main sequence. The star spends most of its time on the main sequence, its horizontal position being determined by the size of the original gas cloud, and produces energy through the fusion of hydrogen into helium. When the hydrogen in the core is nearly exhausted helium becomes the fuel of the nuclear reactions, and this leads to a slow swelling of the star, which takes it along the "red giant branch". This expansion eventually stops and a movement along the "horizontal branch" ensues. At this stage the star is almost exhausted of nuclear fuel and the stellar wind reduces the mass of the star. The outer parts are thrown off, and at the same time the core contracts and forms a small, white dwarf star of high density and high surface temperature that is located at the bottom left corner of the HR-diagram.

The HR-diagram initially served as way to organise and illustrate collected astronomical data, but turned into an important tool for reasoning about the composition and evolution of stars. This is not a formal or symbolic model that provides exact answers, but a conceptual model that aided astronomers in speculating and developing theories about the birth and death of stars.

Mouse Models of Neurofibromatosis

All humans run the risk of being afflicted by diseases that are more or less harmful for our health and well-being. Many diseases, such as common flu or pneumonia, are caused by microbes (bacteria, viruses or parasites) that enter the body and cause the symptoms associated with the disease. But there are other diseases that are not primarily caused by infection, but rather by a combination of environmental and genetic factors. Possibly the most obvious example of this is cancer, which is caused by mutations (random genetic changes) that the cells in our bodies undergo as we age. To put it simply the genetic changes cause the cancer cell to stop listening to the controlling signals in the body and instead divide in an unchecked manner. The risk of contracting cancer of a certain type is not equal for all individuals; it is affected by external factors such as smoking, but also by heritable predispositions. The reason for this is that some people carry mutations inherited from their parents that increase the probability of developing cancer.

There are many other diseases besides cancer that are caused by genetic changes, and among the more common, that are still rare compared to many other diseases, are cystic fibrosis, haemophilia and neurofibromatosis. The latter disease is special since it, in contrast to most other heritable diseases, it is caused by a mutation in a single gene. The disease was first described in 1882 by Friedrich Daniel von Recklinghausen and is therefore also known as von Recklinghausen disease. The symptoms are many and varied, but the most common include superficial neurofibromas (a kind of benign tumour that consists of nerve cells), large birthmarks, skeletal lesions, optic glioma (benign brain tumours) and mental disorders such as learning disability. None of these symptoms are life-threatening in themselves, but together they lead to a lowered average life span and a reduced quality of life.

People who develop the disease carry a mutated version of the gene NF1, that codes for the protein neurofibromin. The disease is considered dominant since it is sufficient for one of the two copies of the gene to be mutated for the disease to appear. Such an individual with one mutated copy is denoted $NF1^{+/-}$, while a healthy person is denoted $NF1^{+/+}$. The precise function of the protein neurofibromin has not been established, but from studies carried out in cell lines we know that it has an inhibitory effect on an intra-cellular signalling pathway known as RAS. The activity of the pathway has in turn been connected to cell division, suggesting that the lack of neurofibromin could lead to increased cell division.

This mechanism suggests a possible explanation as to why the tumour-like symptoms appear, but is far from a satisfactory answer to the question of how the disease develops. How should this problem be approached? One option is to carefully study the course of the disease in patients, but this is often a slow process that in addition is afflicted with a number of practical and ethical issues. Another option would be to study the disease in an organism different from humans, that is easier to handle and manipulate both from a practical and ethical perspective. Such an organism can be viewed as a model of a human and is therefore called a model organism. In many cases scientists have decided to use mice, the reasons being that they are easy to handle, have a short life span, and are relatively similar to humans on a genetic level. In fact 90 % of human genes have a counterpart in mice. However, before a genetically caused disease can be studied in mice one technical challenge needs to be solved: How do we manufacture a mouse with exactly the right set of mutations?

One technique for creating mice with certain mutations, or "knock-out" mice as they are often called, was developed during the 1980s and provided scientists with unprecedented opportunities to investigate the effects of inherited mutations.[19] The technology is based on a naturally occurring mechanism in animal cells known as homologous recombination. This is a process that allows genes to move from one chromosome to another, and occurs spontaneously in our germ cells in order to "shuffle" our genetic material. A gene you have inherited from your mother can in this process end up on a chromosome you have inherited from your father, and vice versa. The other cornerstone in the discovery were so-called embryonal stem cells, which have the ability to transform into any cell type in the body, anything from a white blood cell to a muscle cell.

In the first step of the process a mutated copy of the gene of interest is created and inserted into embryonal stem cells where it is incorporated into the DNA of cells through homologous recombination. The next step is to inject these genetically modified cells into a mouse embryo at the blastocyst stage when it contains roughly 100 cells that form a round mass. The manipulated blastocyst that contains both normal and mutated cells is implanted into the uterus of a female mouse, and the embryo develops into a chimeric mouse that carries a mix of healthy and mutated cells.[20] In the final step of the procedure the chimera is bred with a normal mouse, and some of the resulting offspring will be fully transformed.[21]

$NF^{+/-}$ mice were created for the first time during the 1990s and it turned out that they had similar symptoms to human patients. An important observation was that mice that lack both copies of the gene ($NF^{-/-}$) do not survive the embryonal stage, but die of heart failure before birth. This suggests that NF1 plays a central role in the development of the mice and in the formation of internal organs. $NF^{+/-}$ mice on the other hand develop normally, but show an increased risk of developing cancer and often die from leukaemia or tumours in the adrenal gland. Similarly to humans they also exhibit an increased risk of developing brain tumours and learning disabilities. With the aid of these mice the course of the disease and the development of certain symptoms have been studied in detail, and in addition possible treatments have been tested.

---

[19]The discoverers Capecchi, Evans and Smithies were in 2007 awarded with the Nobel prize in physiology or medicine.

[20]The word derives from an animal in Greek mythology known as the Chimaera ($\chi\iota\mu\alpha\iota\rho\alpha$) that had three heads: one lion, one goat and one snake head.

[21]The chimera will produce germ cells (that just like normal germ cells only contain half the chromosomes) and a fraction of these will contain the mutated gene, while most will be normal. When a germ cell that contains the mutated gene fuses with a normal germ cell a mouse with a single mutated copy of the gene is created, i.e. a $NF^{+/-}$ mouse.

In order to isolate the symptoms connected to the nervous system and in particular the brain, an additional genetic technique has been developed. Instead of creating embryonal stem cells that have a mutated copy of the gene, one can incorporate a so called Cre-Lox construct into the DNA of the cells, which makes sure that the mutated gene only is expressed in nerve cells.[22] Hence it is possible to create a mouse that is $NF^{+/-}$ in nerve cells and $NF^{+/+}$ in all other cells. These mice exhibit symptoms that are very similar to the ones that appear in patients with neurofibromatosis, and this model therefore represents an even sharper tool in the hunt for a better understanding of the disease and novel treatments for these patients.

This brief exposition of mouse models and their application in genetic research is simplified and lacking in detail, but hopefully still provides an insight into how scientists use specially constructed model organisms in order to study complex and multi-faceted diseases such as neurofibromatosis.

Scale Models in Ship Engineering

When something large is to be transported the best option is to go by sea. Although it is much faster to ship goods by air across the globe, the price of sea transport is considerably lower. Historically speaking, waterways were the only trustworthy means of transport even between nearby locations. For example almost all transport in the north of Sweden occurred along the rivers and the coast, and up until the 1950s the road connecting Stockholm and Haparanda on the Finnish border, that runs along the Baltic coast, was no more than a tiny gravel road. The sea has since times immemorial connected cities and brought continents closer, which is evident from old maps where in most cases the area occupied by land is exaggerated compared to that of the sea.

Since prehistoric times, people have probably been aware that the shape of the hull affects the properties of a ship, such as its speed and stability. This knowledge was however not acquired through scientific and systematic studies, but rather by trial-and-error and was passed down through generations. In the chapter that discussed Christopher Polhem (p. 23) we reported on his groundbreaking systematic testing of different shapes of ship hulls. Since then ship builders have, motivated by financial and environmental benefits (unfortunately often in that order), tried to optimise the shape of the hull in order to combine security, loading capacity, speed, construction

---

[22] Also in this case the tools have been borrowed from nature. The Cre gene originally belongs in a so called bacteriophage, a virus that infects bacteria, and the protein it codes for has an ability to cut out pieces of the DNA that are surrounded by a specific genetic sequence. This matching sequence is known as the LoxP and is 34 base pairs long, and is placed on each side of the NF1-gene, flanking the gene of interest. The last detail is to place the Cre gene at a location in the DNA so that it becomes assoaciated with genes that are expressed exclusively in nerve cells. Simply put: in knock-out mice the Cre gene is only activated in nerve cells. The corresponding protein is produced, locates the LoxP sequence and removes it together with the NF1-gene, and the end result is mice that are $NF^{+/-}$ in nerve cells and $NF^{+/+}$ in all other cells.

Fig. 14 A scale model being
tested in the towing tank of
SSPA in Gothenburg. The
purpose of this study was to
estimate the wetted area of
the hull at different speeds.
Reprinted with permission
from Jose Maria
Perez-Macias

efficiency and operational efficiency (unfortunately often not in that order). One
example where speed was prioritised was among the famous clippers that competed
against each other for the quickest transport of fresh tea from China to England. A
more contemporary example is provided by the America's Cup, the most prestigious
sailing competition in which modern ships made out of high tech material and on
their limit of maximal strain and stress race against each other.

Traditionally the shape of a hull is optimised by constructing scale models that
are tested in a towing tank, where the model is dragged through water, and the water
resistance and other characteristics are measured at different speeds. Such a setup
can be seen in Fig. 14, where it was used in order to estimate the area of the hull that
was immersed in water at different speeds.[23] It is also possible to add tiny particles
to the fluid and in such a way follow the movements of the fluid along the hull. This
type of measurement gives rise to flow lines that illustrate how the surrounding fluid
travels past the hull.

When working with scale models the aim is for the model to have three important
properties: geometric similarity (i.e. it should have the same shape as the original),
kinetic similarity (the flow lines should agree with those around the real hull) and
dynamic similarity (the ratio between forces acting at different locations on the hull
should be the same for the model and reality). Geometric similarity is easiest to
achieve, since it is accomplished with a miniaturised version of the original ship, but
this construction does not necessarily satisfy the other two criteria. This is because,
relatively speaking, water becomes more viscous at shorter length scales. For a tiny
insect, swimming in water it is more like swimming in treacle. The so called Reynolds

---

[23]Perez-Macias Martin, J.M. (2009). Estimating wetted area of a model-hull from a set of camera
images using NURBS curves and surfaces. Master's thesis at Chalmers University of Technology.

Fig. 15 This model of a hull was constructed with the application Comsol Multiphysics, which uses the finite-element method to simulate physical processes. The *upper panel* shows the triangulation, i.e. the grid on which the numerical solution is calculated. Please note that the triangles change shape and also become smaller closer to the hull where higher accuracy is required. The *lower panel* shows the flow of water around the hull with the aid of white flow lines and a coloured surface, where brighter colour corresponds to a higher velocity of the water. The grey colour denotes the speed of the vessel itself (20 m/s) and serves as a reference. It is clear that the speed of the water relative to the hull is highest at the bow (at the front) and very low in the turbulent flow behind the stern

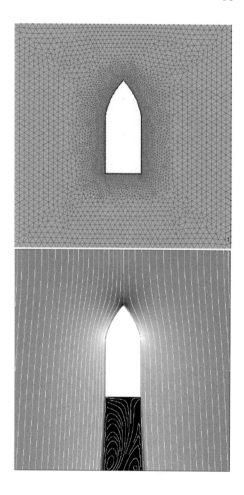

number is a measure of this effect.[24] This implies that neither the kinetic nor the dynamic similarity is preserved for naive scale models, and therefore certain tricks have to be used. One solutions is to use a fluid that has a lower viscosity compared to water, or grooves can be added to the hull of the scale model, creating local turbulence that increases the effective Reynolds number.

Today it also possible to study the properties of ships using computer simulations of Navier-Stokes equations that govern the motion of fluids.[25] When solving Navier-

---

[24]The Reynolds number is a dimensionless quantity that is defined as $\rho V L / \mu$, where ρ is the density, μ is the viscosity and V and L are the "typical" velocity and length scales. A Reynolds number above 2100 corresponds to turbulent flow, while lower numbers indicate laminar flow.

[25]The Frenchman Navier and the Englishman Stokes, independent of each other, derived these equations during the 1820s as a model for the movement of a fluid. The derivation is based on the preservation of the physical quantities: mass, momentum and energy, and is usually expressed in terms of two coupled partial differential equations. With the aid of these equations the fluid is

Stokes equations numerically in two or three dimensions it is common to partition
the computational domain into triangles (or tetrahedrons in the 3D-case), just as in
Fig. 15. On each triangle one starts the simulation with an initial condition, solves
the equations locally and adjusts the solution to the neighbouring triangles, until a
global convergence is achieved. This is known as the finite-element method (FEM)
and is a powerful tool for modelling partial differential equations in complicated
geometries. The lower panel of Fig. 15 shows the result of such a computation,
where the flow around a ship that travels with a speed of 40 knots (roughly 20 m/s)
has been simulated.

Despite the fact that our ability to use computers in order to determine the properties of ships is many times greater than it was 30 years ago, scale models are still a
competitive alternative when it comes to ship building. Since the 1980s scale models
and numerical solutions of Navier-Stokes equations have been used in parallel, but
as computers become more and more powerful we anticipate a shift to occur in the
balance between the two. Most likely computer models and simulations will become
more common, but is unlikely that they will completely outcompete their material
precursors, at least in the foreseeable future.

Epilogue

This concise book sheds light on the concept of scientific models from a number of perspectives. Starting with our human perception and mental world, and moving onto representation and caricatures in the arts, we finally arrive at the scientific concept of a model. We also view models from a historical and philosophical perspective, and discuss not only different kinds of models but also (on the basis of interviews with actual scientists) different approaches to modeling. The picture that emerges is complex and versatile: scientific models represent a broad concept that is used by many people in a variety of contexts. The meaning ascribed to the word varies between different disciplines, and the importance attributed to models has also changed over time.

The concept can be traced from classical mechanics, and into other fields of physics, but is now used in all scientific disciplines, having become so influential that modern science can hardly be imagined without models. This central role is however a fairly recent development accelerated by the rise of computers, which have made simulations of complex models a common approach to solving scientific problems.

We show that modelling can be viewed as an iterative process, and how in that process one must strike a balance between simpler and manageable models and more complex models with higher predictive power. We also point out the dangers that modelling can entail, particularly with regard to policy making and how models are perceived by the public.

Lastly we hope that this brief exposition has done justice to the fascinating and powerful models that are all around us. We hope that reading this book will make it easier for the reader to bridge the gaps between different scientific cultures, and that in the long run it will lead to more and fruitful interdisciplinary collaborations that overcome the traditional boundaries of science.

© Springer International Publishing Switzerland 2016
P. Gerlee and T. Lundh, *Scientific Models*, DOI 10.1007/978-3-319-27081-4

Further Reading

The following books and articles were helpful to us when writing this book and might be of interest for the keen reader.

Books

- Bailer-Jones D (2009) Scientific models in the philosophy of science. Pittsburgh University Press, Pittsburgh
- Cottingham J (ed) (1992) The Cambridge companion to descartes. Cambridge University Press, Cambridge
- Körner TW (1996) The pleasures of counting. Cambridge University Press, Cambridge
- Losee J (1972) A historical introduction to the philosophy of science. Oxford University Press, Oxford
- Pilkey O, Jarvis-Pilkey L (2007) Useless arithmetics. Columbia University Press, Columbia
- Roux S, Garber D (eds) (2013) The mechanization of natural philosophy. Springer, New York
- Suckling CJ, Suckling KE, Suckling CW (1978) Chemistry through models. Cambridge University Press, Cambridge
- Shapin S (1998) The scientific revolution. University of Chicago Press, Chicago
- Wartofsky MW (1979) Models: representation and the scientific understanding. D. Reidel Publishing Company, Dordrecht

Articles

- Achinstein P (1964) Models, analogies, and theories. Philos Sci 31:328–350
- Cartwright N (1997) Models: the blueprints for laws. Philos Sci 64:292–303rr
- Cartwright N, Poidevin RL (1991) Fables and models. Proc Aristot Soc Suppl Vol 65:55–82
- Contessa G (2010) Scientific models and fictional objects. Synthese 172(2): 215–229

© Springer International Publishing Switzerland 2016
P. Gerlee and T. Lundh, *Scientific Models*, DOI 10.1007/978-3-319-27081-4

- Dennett DC (1991) Real patterns. J Philos 88(1):27–51
- Giere RN (2010) An agent-based conception of models and scientific representation. Synthese 172(2):269–281
- Harré R (1959) Metaphor, model and mechanism. Proc Aristot Soc 60:101–122
- Hartmann, S, Frigg, R (2012) Models in science. The stanford encyclopedia of philosophy. http://plato.stanford.edu/entries/models-science/
- Machamer P, Darden L, Craver CF (2000) Thinking about mechanisms. Philos Sci 67(1):1–25
- Rosenblueth A, Wiener N (1945) The role of models in science. Philos Sci 12:316–321
- Wimsatt WC (1987) False models as means to truer theories. Neutral models in biology. Oxford University Press, Oxford, pp 23–55